Jean-François Caron
Violent Alternatives to War

Facing Contemporary Terrorism

Edited by
Jean-François Caron

Volume 1

Jean-François Caron

Violent Alternatives to War

—

Justifying Actions Against Contemporary Terrorism

DE GRUYTER

ISBN 978-3-11-154276-8
e-ISBN (PDF) 978-3-11-072989-4
e-ISBN (EPUB) 978-3-11-073000-5
ISSN 2749-1188
e-ISSN 2749-1196

Library of Congress Control Number: 2021943964

Bibliographic information published by the Deutsche Nationalbibliothek
The Deutsche Nationalbibliothek lists this publication in the Deutsche
Nationalbibliografie; detailed bibliographic data are available on the Internet at http://
dnb.dnb.de.

www.degruyter.com

Acknowledgements

I would like to acknowledge the invaluable input of people who have commented early drafts of this project, as well as the help of my research assistant Aigerim Zholdas who has helped me find some of the primary sources used in this book.

I dedicate this book to the innocent victims of the 9/11 attacks and those who have been killed during the 'War on Terror' ever since that tragic morning I witnessed in horror as a young Parliamentary Assistant in my office at the Canadian House of Commons.

https://doi.org/10.1515/9783110729894-001

Contents

Introduction

As Voltaire (1764) wrote in his *Philosophical Dictionary*, 'war is an inevitable scourge' made worst by the fact that 'murderous war is so much the dreadful lot of man'. Faced with this ineluctable fate, it is perhaps why human societies have always emphasised the importance of making war a moral reality by finding ways to minimise as much as possible the harm that ought to be allowed during warfare. It is from this perspective that war ethicist Brian Orend (2013, p. 9) wrote that 'Almost all major civilizations – from the ancient Egyptians to the Aztecs, from Babylon to India, from China to medieval Europe and contemporary America – have featured quite colourful beliefs about acceptable reasons for going to war, and permissible means of fighting it'. The main intention of those who have thought about these questions, which could be summed up today as being key components of the 'Just War Theory', has always been to find ways to limit the right to resort to violence and the scope of legitimate harm during wartime only to those who are deemed to have lost their immunity against death. This was reflected in many traditions, such as the Indian culture that came up with a written code of conduct for war – the Code of Manu – which prevented combatants from killing fleeing enemies, and those who were sleeping or prisoners of war. The Greeks and the Romans also emphasised the importance of fighting wars with moderation. In this context, Cicero talked about the importance of restraint by not harming a sleeping or wounded combatant, refraining from attacking religious temples, and keeping one's promises. Nowadays, the moral limitations on the legitimate use of violence are bounded by the principles of *jus ad bellum*, while the ethics of fighting well are enshrined within the principle of *jus in bello*.

However, as most pacifists would argue, history has shown that there has always been a sharp difference between what wars ought to be and what they have looked like and, as such, hoping to humanize military conflicts is no more than wishful thinking. Indeed, despite combatants' best intentions and efforts to abide by the moral rules of warfare, wars and military interventions almost always result in their violations. This is why it is largely believed – with reason – that war should always be a last resort, when all non-violent alternatives to war (NVATW) have proved ineffective. According to James Pattison's theory, states have at their disposal a vast array of non-violent options that do not imply the resort to arms, physical harm, or killing, and that can be as effective as any kinetic military action (2018). Among them are non-violent resistance, the imposition of economic sanctions or arms embargoes, and the use of diplomacy. In many cases, these measures, which can also be labelled 'soft war' mea-

https://doi.org/10.1515/9783110729894-002

sures (Gross, 2015; Gross & Meisels, 2017), may lead to the desired outcomes without having to resort to extensive military operations. When such an end result can be reasonably entertained through these measures, they should naturally be privileged by states. From a moral perspective, this idea is intimately connected with those of some of the first philosophers to speak of what is now known as the Just War Theory – lethal actions are permissible only insofar as non-violent alternatives are unavailable or likely to be ineffective at countering the threat (Aquinas, *Summa Theologiae*, II-II, 64, 1; Grotius, *De Jure Belli ac Pacis*, 2.1..4.2)

When we take a look back at the way Western states have fought terrorist organizations in the last 20 years, it is difficult not to think that these alternatives to war might have been more ethical than the decisions to invade Afghanistan and Iraq respectively in 2001 and 2003. In fact, as these two cases have shown, you do not need to be a full-time political scientist to know that relying on full-scale wars to fight the terrorist threat can lead to the worst outcomes as it is virtually impossible to determine with precision the scale of the humanitarian and political consequences of a military intervention. These cases speak for themselves as they have both led to the deaths of tens of thousands of innocent civilians, which is highly paradoxical in light of the logic that supported these interventions. Indeed, if the objective of these wars was to prevent the unjustified death of non-combatants, it is clear that they have simply transferred that risk (Shaw, 2005) to the Iraqi and Afghan civilians,[1] as if 'our' lives had more value than 'their' lives: a premise no reasonable human being would dare defending. In this perspective, it is not difficult to understand why the fight against terrorism has been labelled by some as being terroristic as well.

Moreover, full-scale wars against terrorist threats have resulted in a lack of political stability as well as an incapacity to eradicate the menaces that led to their beginning. In the case of Afghanistan, the projected withdrawal of US forces from the country by the end of August 2021 and the negotiations with the Taliban can be interpreted as a rotten compromise and what Henry Kissinger calls a 'decent interval', which will most likely pave the way for the return to

---

1 As written by journalist Philip Bump (2018), 'During the war [in Iraq] and during the Islamic State militant group's occupation of as much as a third of the country in recent years, the number of deaths runs into the hundreds of thousands, including civilians killed as a result of violence and, more broadly, those who died because of the collapse of infrastructure and services in Iraq resulting from the ongoing conflict'. For its part, the Watson Institute (2020) estimated in October 2019 that a little more than 150,000 people have been killed in the Afghanistan war since 2001 and that 43,000 of them were civilians. In comparison, the 9/11 attacks costed the lives of 2,977 innocent civilians.

power of those who once provided safe haven to members of Al-Qaeda in the years that preceded 9/11 (Caron, 2015). In some cases, these interventions have even been directly linked with the emergence of new threats, as was the case with the Islamic State following the 2003 invasion of Iraq. As we say in French, these interventions have led the US and its allies to undress Saint Peter in order to dress Saint Paul. At the end of the day, nothing has really changed, and it is hard to claim victory when the eradication of the initial threat has led to the creation of another one as – and sometimes more – dangerous than the former. For all these reasons, if these are the inherent disproportionate consequences of waging war against terrorism, there would be serious grounds to argue that this way of dealing with such groups is morally questionable and that we ought to consider alternative options.

However, as attractive as they may sound, there are reasons to believe that the previously mentioned NVATW are not as effective against contemporary terrorist organizations as they are against state entities. Owing to the stateless nature of terrorist organizations, these alternatives to political violence do not have the same effects as when they are used against states. When it comes to state actors, there are reasons to believe that they may change their behaviour through non-violent methods that can involve forms of political or economic pressure and coercion. For instance, the use of what has been called 'coercive diplomacy' (George, 1991), which can imply sporting sanctions, threats to withdraw political and aid support, or the threat of not going forward with promised economic and political aid, can achieve this goal without resorting to war. Here, the reluctance to use the sword against state actors is justified by the presence and proven effectiveness of these non-violent options (Coady, 2008, p. 91). However, these 'carrots' remain largely ineffective against actors like terrorist groups, who could not care less about the possibility of losing diplomatic recognition or promised aid, or being banned from participating in the Olympics or FIFA World Cup, which is why the previously mentioned intermediary non-violent options cannot always be entertained against these groups.

In light of this situation, it seems as though we are back to square one: either states are to wait until they are the victim of an attack or they ought to wage wars against terrorist organizations threatening them. Both options are unfortunately problematic, and this book will discuss. Indeed, resorting to an entirely reactive strategy that would imply having the right to resort to violence only once a state has suffered a terrorist attack would not satisfy its obligation to defend and protect its citizens. Needless to say, this is difficult to justify as 'the potential for mass-casualty terrorism renders a reactive strategy imprudent at best and potentially immoral' (Bellamy, 2006, p. 163). However, and for the reasons previously evoked, relying on NVATW against such groups is also questionable as they may prove ineffective against them. In return, resorting to war against terrorist organizations or states harbouring them may not only be inefficient at eliminating the menace but could also be detrimental to states' moral obligation to limit the impact of violence on civilians. This, therefore, begs the question of what states should do when facing a terrorist threat. Obviously, this paradox shows how the international community is now at a turning point when it comes to what ought to be done against such a threat, which implies a different view of what is an acceptable degree of lethal force against such groups (Fisk & Ramos, 2016, p. 2).

As this book will argue, there is a need to think of alternatives to war that will imply the legitimisation of proactive measures allowing states to effectively prevent terrorist attacks through the use of limited kinetic force. This would serve the purpose of avoiding the terrible and unpredictable consequences of warfare and cementing them as 'measures short of being acts of war'. As these types of actions will require the use of direct military action, they fall outside the realm of the previously discussed NVATW. However, given their limited nature in terms of scale and time, they should not be perceived as acts of war but rather as alternatives to war. This is why they will be called 'violent alternatives to war' (VATW).

VATW are opening up a space in the binary opposition between 'hard war' and 'soft war' (Gross & Meisels, 2017). In other words, measures that fall somewhere between the use of bombs, bullets, and missiles on a massive scale and non-kinetic actions (such as economic sanctions, non-violent resistance, and diplomacy), and that involve resorting to a limited amount of violence that can encompass death and destruction but would nonetheless fall short of being considered acts of war, can be justified in the fight against contemporary terrorist organizations. These measures are also thought to establish a well-needed balance between two fundamental rights: the right to life of those threatened by terrorists (which will require defensive measures on the part of states) and the obligation to limit violence only against those who have lost their immunity against being harmed. As it stands right now, not acting against terrorist groups threatens the former right, while waging full-scale wars against them will result in violating the latter by creating mass civilian casualties. Both options are morally unacceptable.

Do VATW Justify a New Category in the Just War Theory?

The question of how to achieve more strategically efficient methods of fighting terrorism in a way where the ethical principles of warfare are respected, has been a highly debated topic over the last 20 years. Some have suggested the need to create an additional category to the Just War Theory known as *jus ad vim*, a notion referring to the just use of force. Mentioned for the first time by Walzer in the fourth edition of his seminal book, *Just and Unjust Wars* (2006), *jus ad vim* is seen as a theory of just and unjust uses of force that is separate from the *jus ad bellum* category. For Walzer, while the latter notion refers to war itself, the former is a category that includes all forces short of war, whether they be non-violent ones as discussed by Pattison, or VATW, namely the establishment of an inspection system intended to monitor a country's management of its current stock of weapons, or its capacity to develop and acquire new ones, economic or military embargoes, or the imposition of no-fly zones (2006, p. xiv). According to Walzer, this kind of containment system, implemented against Iraq from 1991 until 2003, proved highly efficient in its preventing Saddam Hussein from acquiring and developing any weapons of mass destruction (WMD) which he might have used either directly, or by supplying them to terrorist organizations, and from harming the civilian population in the northern and southern parts of the country, such as the Kurds and the Shiite Muslims. For Walzer, these measures short of war proved themselves to be highly efficient and made the 2003 invasion of the country by the US and its allies absolutely unnecessary. As he writes:

> The harsh containment system imposed on Iraq after the first Gulf War was an experiment in responding differently [to a threat]. Containment had three elements: the first was an embargo intended to prevent the importation of arms (which also affected supplies of food and medicine though it should have been possible to design a "smarter" set of sanctions). The second element was an inspection system organized by the UN to block the domestic development of weapons of mass destruction. The third element was the establishment of "no-fly" zones in the northern and southern parts of the country so that Iraq's air power could not be used against its own people. The containment system was, as we now know, highly effective. At least, it was effective in one sense: it prevented both weapons development and mass murder and therefore made the war of 2003 unnecessary (Walzer, 2006, pp. xiii-xiv).

The ability to prevent an aggression or mass murder, without having to resort to war, also comes with a highly valuable moral advantage. Indeed, we need to acknowledge that despite our best intentions and our genuine attempts to respect the moral rules of warfare, full-scale conflicts always have tragic consequences for civilians, as has been the case in Afghanistan, Iraq, and more recently, Libya. Alongside the previously mentioned unforeseen and uncontrollable effects of full-scale wars, the use of violence on a large scale will always lead to the violation of the moral rules of warfare; more specifically, the principle of discrimination between combatants and non-combatants. This can be the outcome of various factors. Out of stress, fatigue, or a desire for vengeance, soldiers may end up targeting civilians. In an emergency situation, soldiers may confuse a civilian building for a legitimate military target. Despite their sophistication, bombs and missiles can always malfunction, landing in an unexpected environment. Although such risks may be minimal, it grows exponentially when the numbers run into the thousands. Therefore, we need to entertain the possibility that these unfortunate consequences may be prevented if we were to allow the use of measures associated with the *jus ad vim* logic that may effectively prevent terrorist organizations from attacking states in a way that would also prevent these sad consequences of war. In fact, we might even say that measures short of war are morally preferable to acts of war and must, consequently, be used whenever they are feasible.

Daniel Brunstetter and Megan Braun (2013) are probably those who have most fiercely defended the importance of creating the separate category of *jus ad vim* for measures short of war that fall within our notion of VATW. This is on the basis that, unlike warlike actions, the scale of violence they imply is much more limited, such as drone strikes that have a scale and effect exponentially inferior to that of resorting to full-blown wars. Furthermore, by allowing states to protect themselves through these limited actions, it is believed that such measures will prevent the risk of escalation to full-blown war[2]. Considering the tragic consequences of the 2003 invasion of Iraq, it is hard not to agree with this account of the effectiveness of these measures in their capacity to achieve an objective without having to resort to war and its atrocious humanitarian consequences.

This idea has nonetheless led to fierce debates between just war theorists. For some, like Helen Frowe (2016), creating this new category would be redundant as she believes that the logic and criteria supporting them are not distinctive from the *jus ad bellum* category. Indeed, this category includes all the criteria and measures that ought to be considered before engaging in a rightful war. Among these are the idea that war ought to be a last resort and that all other peaceful options that might prevent this sad reality from occurring ought to be tried first (or at least those that have a reasonable chance of succeeding) (Aloyo, 2015). This is based on the understanding that the Just War Theory can rightfully be labelled as an approach that has a conditional view of peace, mean-

---

2 Brunstetter and Braun argue that '(...) jus ad vim should not be conceived of as part of the actions leading up to war, but rather should serve as an alternative set of options to the large quantum of force associated with war' (2013, p. 97) and that 'If engaging in jus ad vim actions has a high probability of resulting in war, then one could argue that such actions are not justifiable, and must be subject to the stricter jus ad bellum regime' (2013, p. 99).

ing that only extreme circumstances that cannot be prevented by peaceful means can justify war. Accordingly, this version of pacifism – unlike the one defended by absolute pacifists like Gandhi or Martin Luther King – which is an inherent component of the Just War Theory, is not a synonym of passivism. From this perspective, the *jus ad bellum* category would be constituted of various measures designed to prevent states from going to war. Consequently, since NVATW and VATW are supposed to fulfil that objective, it is logical to count them as constitutive measures of the *jus ad bellum* category as well. In this sense, *jus ad bellum* must be considered as a spectrum of measures ranging from naming and shaming, diplomatic and economic sanctions, arms embargoes, to more violent measures such as targeted killings. According to this view, thinking about measures that might prevent the need to resort to full-scale war is nothing new and is an already widely accepted idea among state actors whose decision to engage in or to abstain from waging war always results from an assessment of the merits and demerits of the various alternatives to war (whether non-violent or violent).

These arguments are rather convincing, but I feel they do not allow for a clear and precise understanding of what VATW are all about. Conflating them as philosopher Michael Walzer has done with NVATW as part of a set of actions located on a continuum that can ultimately lead to war is problematic, because VATW are fundamentally different from NVATW and do not imply the same moral challenges. In fact, not differentiating between them is a mistake usually made by Just War scholars who do not make a necessary distinction between non-violent and violent force, as though trying to influence an actor to change his or her course of actions through different types of non-violent means – whether it is through diplomacy or economic sanctions – is the same as eliminating him. Of course, both sets of measures serve the same objective – to paraphrase von Clausewitz – in preventing the resort to full-scale war by compelling our enemy to fulfill our will. However, from a moral perspective, there is a huge distinction to be made between pressuring of an entity to change its original course of action through non-violent actions, and the destruction of its infrastructure through bombing raids. In fact, contrary to the former set of measures, the latter implies an exception to the natural rights we ought to enjoy as human beings by legitimizing the destruction of property and the loss of immunity from harm. When such violations ought to be justified, the degree of danger posed by these individuals is obviously of a more serious nature than the one posed by individuals whose harm to others is thought to be preventable through non-violent pressures. One cannot simply think about the resort to these measures as being equivalent to the same moral reality, as one implies the clear and explicit violation of a fundamental natural right. For instance, although both the death penalty and speeding fines are instruments of law enforcement mechanisms in some societies, the former is perceived with more moral skepticism than the threat of the latter. This is why killing and causing destruction – even on a limited scale – is not a form of pressure that can be equated with NVATW. This calls for NVATW and VATW to be treated differently. If the legitimacy of the former can be assessed within the *jus ad bellum* category, the latter set of measures will require a different moral evaluation by developing a novel understanding of what constitutes an imminent threat, and about how the killing of individuals in a situation where there is no ongoing war between states can be justified – requirements that are currently not constitutive elements of international law and the moral rules of warfare. This is obviously a major shortcoming of the Just War Theory, which has been conceptualized and developed around the idea of inter-state wars. Indeed, nowadays, these sorts of wars are the exception rather than the norm.[3]

As such, NVATW and VATW are morally distinctive, and justifying the latter actions will require a separate discussion about the use of force and the legitimacy of killing. This is a feature which Walzer, unfortunately, does not consider in his assessment of *jus ad vim*, which has led to Frowe's previously mentioned criticisms against the necessity of this notion forming a new category within the Just War Theory because, as she rightly points out, most of the actions discussed by Walzer – namely embargoes, an internationally enforced inspection system or the imposition of no-fly zones – are undoubtedly NVATW and already thought to be non-lethal constitutive elements of the *jus ad bellum* category. However, and as I have already discussed, the problem is that what is referred to as measures short of war are not solely restricted to NVATW, but also to lethal actions. Being fundamentally different from NVATW, does this mean that thinking about VATW justifies the creation of a fourth category within the Just War Theory? In other words, while the former set of measures would fall within the category of *jus ad bellum*, the latter would rather necessitate the inclusion of an additional category because they imply a resort to lethal actions. I believe this is a diversion from what ought to be the real question that needs to be addressed, namely how VATW as a specific type of force short of war can be morally justified. My goal in this book is to explain how these moral questions, raised by the necessity for envisioning VATW against contemporary terrorist organizations, can be addressed. At the end of the day, whether VATW are to be considered as part of the *jus ad bellum* category, or of a newly formed *jus ad vim* category, is for me a simple matter of rhetoric, since in both cases they would not be considered

---

3 Out of the 118 conflicts that took place between 1989 and 2004, only seven were interstate wars (Harbom & Wallensteen, 2005).

as *casus belli* but as actions short of being acts of war, which is in my view what ultimately matters the most.

What Will Be Discussed

Arguing that the fight against international terrorism will necessitate some sort of military response (Keohane, 2002, p. 29) through the extension of allowed violence but not to a point where we might end up justifying full-scale wars[4] is a very tricky task that requires an innovative approach. The main problem is that allowing states to resort to VATW implies facilitating a more generous use of force in a way that should not at the same time be an overly tolerant approach that justifies violence in all its forms. We need to be clear that there is an acceptable level of violence that can be used to deter terrorists and that there should be a threshold that VATW should not cross, otherwise they would become akin to full-scale wars: a reality that needs to be avoided. The question is to determine the limit after which a forceful action ought to be considered an act of war. Similarly, the identity of those against whom these measures should be used also needs to be delimited very clearly. First, even though terrorist groups ought to be the prime targets of these measures, a number of criteria need to be satisfied before striking against them can be legitimised. Determining the identity of who should be victim to a legitimate form of violence is a fundamental moral question. After all, justifying political violence – especially measures that may prove lethal to some individuals – comes with a significant challenge, namely the permissibility of killing in a context where states are not at war. If it is possible to justify the loss of immunity after the start of a war, it is a rather more difficult task to do so when two entities are not at war. Justifying someone's death when that person has not already attacked you is not a matter that should be taken lightly, and this is why the permissibility of killing through the use of VATW needs to be grounded in a strong and convincing justification of this form of violence.

In order to fully understand the potential of VATW against contemporary terrorism, it is first necessary to understand the constitutive elements of this form of political violence. This assessment will then allow us to understand the dilemma states targeted by a terrorist threat are facing. Indeed, there are reasons to be-

---

4 A necessity echoed by Brunstetter and Braun, for whom 'the evolving nature of threat [such as terrorism and the use of weapons of mass destruction] illustrates the need for a more calibrated view of force, (...) but not necessarily the declaration of a full-scale international war' (2013, p. 93).

lieve that NVATW will not be effective against these organizations, which will call for the resort to anticipatory violence before their menace is fulfilled. The apparent solution in this regard is resorting to a pre-emptive attack, which is already a constitutive element of the *jus ad bellum* category and a lawful use of force. However, because the right to strike pre-emptively against an enemy has been established in a period when conflicts were between state entities, this principle is currently largely inapplicable against terrorist groups. Owing to threats from de-territorialized terrorist organizations that are attacking covertly and without precursory signs, the criterion that usually allow states to rightfully resort to pre-emptive actions hardly apply against this type of danger. As a result, this shortcoming creates a situation where states are left with few or no effective options at their disposal to face this threat and to protect their citizens. This assessment will then lead in the third chapter to a thorough discussion about how the notion of pre-emption ought to be rethought in light of this new threat in order to provide states with alternatives that will allow them to actively prevent these groups from attacking. It is, of course, one thing to argue that states have the right to respond in an anticipatory manner to threats made by terrorist organizations, but it does not inform us at all on the degree of force they ought to use against them. This is what the fourth chapter will analyse by focusing on the various forms of VATW and the moral rules that must limit their use. One thing is, however, for sure: since one of their main objectives is to avoid having to resort to full-scale wars, they will have to be violent actions of which the scale and effects are limited. As it will be shown, these measures are by no means a novelty, as some are very similar to the interpretations by international lawyers in the 19th century, namely, 'neither wholly warlike nor wholly peaceful', to use the expression of T.J. Lawrence (quoted in Neff, 2005, p. 231). During that period, it was generally admitted that states were allowed to resort to force on a limited scale in response to an offence that did not justify the recourse to a full-scale war following unsuccessful attempts to obtain satisfaction or reparation through peaceful means. They are, in this sense, very different from actions scholars like George Kennan have called 'measures short of war' and that are more akin to a hegemon's desire to prevent a rising power from shifting the balance of power. Lastly, as already mentioned, justifying political violence – especially measures that may prove lethal to some individuals – comes with serious moral and political problems. One is the justification of the permissibility of killing in a context where states are not at war. According to the dominant tradition, the loss of immunity from being harmed affects only soldiers, irrespective of the side they fight on, because they all have the potential to harm other combatants by joining their country's armed forces. As Walzer puts it, 'soldiers as a class are set apart from the world of peaceful activity; they are trained to fight, provided with weapons, required to fight on command. No doubt, they do not always fight; nor is war their personal enterprise. But it is the enterprise of their class, and this fact radically distinguishes the individual soldier from the civilians he leaves behind' (2006, p. 144). On the basis that non-combatants are by definition unarmed and do not pose a threat to anyone, killing them is forbidden (as it would be akin to murder and not self-defence). But this loss of immunity applies only once a war has started. As a consequence, there is a need to analyse how it might be possible from a moral perspective to justify the killing of terrorists prior to the existence of a state of war between them and a state. This is what the fifth chapter will discuss.

My hope is that this theory will enable policymakers to have a better view of what can legitimately be done to protect their states and citizens from the murderous actions of terrorist organizations. Owing to the inability of the traditional Just War Theory to provide a satisfactory solution to this contemporary challenge, there is a need to go off the beaten track by rethinking our understanding of violence. If we cannot deny that launching a full-scale attack with the help of tanks, bombers, fighter jets, and thousands of infantrymen against an unsuspecting state constitutes an act of war, we cannot say the same when a training camp of a terrorist organizations that has pledged to kill as many citizens as possible in Paris, London, or New York is being targeted in the middle of Afghanistan by a drone strike or by a surprise airborne mission by elite troops. If the former ought to be considered an unjustified act of violence, the latter must – when certain criteria are met – be considered a legitimate use of violence that provides an effective alternative to the destruction of war. It is only through this subtle conceptualisation of violence that we may be able to avoid situations similar to the ones witnessed in Afghanistan and Iraq. If the intentions were noble – of course, more in the former case than in the latter – the methods used turned out to be catastrophic as they left the regions in a profound disarray and led to the emergence of new terrorist threats and the death of hundreds of thousands of innocent civilians.

Chapter 1:
Understanding Contemporary Terrorism

According to international law, there are two ways of understanding violence: either as a legitimate action when used for self-defence or an illegitimate action when used to unjustly attack another political entity. Following this approach, states can rightfully resort to violence only when they have a just cause, which can only be the case when their sovereignty has already been violated or when it is about to be violated by another state. However, this conceptualisation of legitimate violence that emerged from the Second World War is now problematic and has demonstrated its incapacity to successfully meet the challenges of contemporary conflicts (Gross, 2010). The quintessential example of this shortcoming was the inability of the international community to stop the mass violation of human rights in Rwanda and Bosnia and Herzegovina in the 1990s. These terrible crises led the international community to enlarge its notion of legitimate violence by creating the Responsibility to Protect (R2P) principle, which now allows the violation of another state's sovereignty in the case of mass murders of civilians.

The complexities of contemporary terrorism have also showed the deficiencies of the post-1945 Westphalian order. The problem is two-fold. Indeed, it is very difficult – if not impossible – for states to foresee when these groups will attack them, which increases the vulnerability of their citizens. In this regard, one approach has been the waging of preventive wars against entities suspected of collaborating with terrorist organizations. The 2003 invasion of Iraq, which had numerous tragic consequences, namely political instability in the region, civilian deaths, and an incapacity to fully eliminate the terrorist threat, is the best example in this regard. At the other end of the spectrum, once states have suffered an attack, their reaction has been to wage a full-scale war against their enemy. The consequences of such actions, namely in Afghanistan, have been as tragic as those in post-war Iraq. There is, therefore, an obvious need to find a more efficient way of facing this menace in a manner that will simultaneously prevent innocent civilians from being killed by these organizations and avoid the devastating effects of a full-scale war on civilians abroad. In other words, as was the case following the mass violations of human rights in the 1990s, we must revise the concept of legitimate use of force against terrorist organizations.

But the first challenge is obviously to define terrorism. Although scholars have debated this question for nearly 50 years, no widely accepted definition

https://doi.org/10.1515/9783110729894-003

of this form of political violence has emerged[5]. In this sense, I am totally aware that my definition of this phenomenon will be contested by many who might see it as too limiting or as an attempt to justify some forms of political violence that I will refer to under the umbrella of 'guerrilla warfare'[6]. I nonetheless believe that my definition allows for the establishment of a difference between the so-called 'old' and 'new' forms of terror in a way that helps explain the shortcomings of the moral criteria that ought to justify violence against contemporary groups such as Al-Qaeda or the Islamic State.

1.1 Terrorism and the Indiscriminate Use of Violence

A lot has been written about the nature of modern terrorism, but many of these works have not been able to specify its true nature. More precisely, its main characterizations are usually thought to be the large number of victims, transnational actions, and the sacred role played by religion that leads to the rationalization of even the most immoral actions (Rapoport, 1984). However, such an assessment of the uniqueness of this contemporary form of violence is misleading. Firstly, emphasizing the role played by religion in contemporary terrorism neglects the fact that it has been a key factor in the mass murder of individuals for centuries. Indeed, it has been estimated that the Thugs – whose actions were designed to please Kali, the Hindu goddess of terror and destruction[7], as a way to keep the universe in balance (Sedgwick, 2004, p. 798) and who operated largely in present-day India – killed an estimated 500,000 to one million individuals during the last three centuries of their history[8]: a conservative estimate that is

---

5 For Walter Laqueur, because terrorism has taken so many forms in so many contexts, it is impossible to come up with a comprehensive definition (1977, p. 5). Authors Leonard Weinberg, Ami Pedahzur, and Sivan Hirsch-Hoeflers' reviews of journal articles from 1977 to 2001 have yielded a total of 73 definitions of terrorism (2004, p. 780).

6 This distinction between terrorist groups and guerrilla groups is an implicit understanding in a lot of studies on political violence (Carter, 2016, p. 133).

7 As David C. Rapoport writes, "The Thug considers the persons murdered precisely in the light of victims offered up to the Goddess, and he remembers them, as a Priest of Jupiter remembered the oxen and as a Priest of Saturn the children sacrificed upon the altars" (1984, p. 664).

8 Rapoport writes that "No one knows exactly when the Thugs (often called Phansigars or stranglers) first appeared. Few now believe that the ancient Sagartians, whom Herodotus describes as stranglers serving in the Persian army, are the people whom the British encountered in India some 2500 years later. But there is evidence that Thugs existed in the seventh century, and almost all scholars agree that they were vigorous in the thirteenth, which means that the group persisted for at least six hundred years" (p. 661).

beyond the combined number of people killed by contemporary terrorist organizations. Secondly, if Al-Qaeda, the Islamic State, and the Islamic Jihad are undoubtedly animated by unnegotiable religious dogma and by the transnational nature of their actions, the same could have been said about the Thugs, the Assassins, or the Zealots-Sicarii. More precisely, the Thugs and the Assassins moved constantly across borders, while all three of these movements were driven by an uncompromising willingness to achieve their religious goals. Therefore, if there is a difference between old and new forms of terrorism, it must be found elsewhere: religion is not in itself a novel aspect of terrorism.

There is no doubt that the 9/11 terrorist attacks have been for many people one of their most traumatising memories and these images that many of us still have in mind should provide us with an understanding of what terrorism refers to. Witnessing these planes crash live on TV into the Twin Towers of the World Trade Center and some of those trapped inside jumping out of windows to escape the terrible blaze that resulted from the 10,000 gallons of jet fuel scattered inside the buildings by American Airlines Flight 11 and United Airlines Flight 175 was a haunting experience. If it was not already the case, terrorism became in many people's minds a synonym of total war and of a profound lack of respect for human life. It must, however, be mentioned that this means of warfare by some groups is unique in light of the long history of terrorism. Indeed, most of the groups described as 'terrorist' have been very respectful of the principle of discrimination between non-combatants and the people they believed were legitimate targets. This is no longer the case. For Walter Laqueur, the preeminent historian of terrorism, this way of fighting constitutes a revolution in terrorism. If their predecessors were rational and calculated in their use of violence, the perpetrators of this new form of terrorism know no limits in terms of who deserves to die and the sorts of weapons to be used (Laqueur, 1999). As such, this leads to two major consequences. Firstly, if terrorism refers to these sorts of actions and that they represent a shift with those that we have witnessed in the past from individuals and groups that have been labelled as terrorists, there is a need to refer to these latter actors with a different word. Secondly, this shift in the means of warfare used by some terrorist organizations is the fundamental reason why we need to rethink the legitimate use of force against them as it impairs states' most basic obligation – to protect their citizens from harm.

But before going further, we first need to talk about one major misconception people have about terrorism. As argued by Gérard Chaliand and Arnaud Blin, the terrorist phenomenon is a complex one defined through the lens of an ideological filter that has tended to make it the exclusive practice of non-state actors (2015, pp. 13–26) who use indiscriminate means of violence (Primoratz, 2013, p. 15). This is why it is usually considered an immoral technique that is solely used by rogue insurgents. This may explain why many would be tempted to define terrorism as a political method used by non-state actors to force states to pursue a course of action they would otherwise not follow. This is especially true with the 1983 definition of terrorism by the US Department of State which reads as follows: "terrorism is premediated, politically motivated violence perpetrated against noncombatant targets by sub-national groups or clandestine agents, usually intended to influence an audience" (US Department of State). This definition, which has been privileged by the international community, unfortunately cannot capture the true essence of terrorism. Indeed, this definition is based on a double standard regarding the use of violence, deemed legitimate when used by state actors but terrorist actions when used by non-state actors. Undoubtedly, this definition has a rhetorical purpose, as it allows states to demonise those they are fighting as immoral terrorists which, consequently, may also be a way to justify all actions against them. This double standard is not only problematic because of its potential consequences regarding the ways we fight these groups but also because it is fundamentally untrue, since indiscriminate use of violence can also be a tactic used by states.

In fact, it is a mistake to think of terrorism and its indiscriminate use of violence only as the purview of non-state actors[9]. In history, what Chaliand and Blin (2015, p. 21) have labelled 'top-down terrorism (*terrorisme d'en haut*)' has actually led to exponentially more deaths than 'grassroots terrorism (*terrorisme d'en bas*)'. If it is possible to talk about state-sponsored terrorism (as was the case with Libya under Gaddafi or Iran, which have directly or indirectly engaged in violent actions against citizens or other states), we should not neglect the fact that totalitarian states have made terrorism a core feature of their domestic politics[10]. It is, therefore, unsurprising that the word 'terrorism' derives from the French word *terreur* and refers to the period from 1793 to 1794, when the Jacobins

---

9 This semantic is by no means an accident, since it is in the interest of states to limit the definition of terrorism to non-state actors as a way to delegitimise the actions of the latter while justifying similar measures they choose to implement against them.

10 This is in line with Timothy Shanahan, who mentions the two dimensions of state terrorism. He writes: '"State terrorism" is terrorism carried out by the state's agents acting *as* agents of the state. It follows from this definition that an act of terrorism carried out by a government employee for purely personal reasons would not be an act of *state* terrorism. Likewise, a private individual who carries out an act of terrorism because he believes doing so will benefit the state, but whose action is in no way sanctioned by state authorities, would not have committed an act of state terrorism. State terrorism could be further distinguished into "internal" and "external" state terrorism. The former would be terrorism directed against members of the same state; the latter would be terrorism directed against other states or their members' (2009, p. 195).

imposed on France a regime of systematic terror that spared no one. To be more precise, the word was openly used by members of the *Convention nationale* who declared terror 'an order of business (*à l'ordre du jour*)' on 5th September 1793, which was followed a few days later by the adoption of the Law of Suspects that allowed the state to arrest anyone based on broad and imprecise criteria[11]. Saint-Just, one of Robespierre's fiercest henchman, provided the basis of arrests in a speech on 10th October in which he said:

> We will not have prosperity as long as the last enemy of freedom will breathe. You must punish not only the traitors, but also those who will not be bothered; you must punish anyone who is passive in the Republic and who does not do anything for her. Ever since the French people has decided to show its will, everything opposed to it ought to be considered as its enemy. We must govern with an iron fist and we must not hesitate to oppress the tyrants (quoted in Chaliand & Blin, 2015, pp. 146–147).

Shortly after came the establishment of revolutionary tribunals, that were not bound by any procedural rules as they were able to sentence individuals unable to prove their innocence against the charges brought against them. The fact that thousands of people – a very vast majority of them being totally innocent – were sent to the guillotine shows the terroristic nature of Robespierre's regime. However, the revolutionary tribunals also had another purpose, namely to spread fear throughout French society. Igor Primoratz writes in this regard:

> (...) the trials and executions were also meant to strike terror in the hearts of all those in the public at large who lacked civic virtue, and in that way coerce compliance with, and indeed active support of, revolutionary laws and policies. (...) That is what makes the "Reign of Terror" a case of state terrorism. Jacobins believed such terrorism was a necessary means of consolidating the new regime. As Robespierre put it, terror was "an emanation of virtue"; without it, virtue remained impotent. Therefore the Jacobins applied the term to their own actions and policies quite unabashedly, without any negative connotations (Primoratz, 2013, p. 33).

This logic prefigured what was to happen in the 20th century in all totalitarian states that used similar methods. For instance, Lenin did not hesitate to order his subordinates to resort to the use of indiscriminate violence against the civilian population[12]. Similarly, many were tried and sentenced by puppet tribunals

---

11 In fact, the *commune de Paris* was famous for its definition of who was to be considered a suspect. It wrote in October 1793 that: 'Those who have done nothing against freedom, also have done nothing for it'.

12 For instance, on 9th August 1918, he instructed the Soviet of Nizhni-Novgorod to act in the following way: 'It is obvious that a white-guardist uprising is being prepared in Nizhni. You

under imaginary accusations in Nazi Germany. These two systems of terror were reinforced by the presence of a powerful and arbitrary secret police that had the right to arrest individuals for any reason, which resulted in the destruction of social capital. This is why Hannah Arendt (1958, p. 464) came to the conclusion that 'terror is the essence of totalitarian domination' and why Carl J. Friedrich and Zbigniew Brzezinski (1965, p. 169) wrote that 'total fear reigns' in these regimes.

Moreover, state terrorism has been the appanage of democratic states that have not hesitated to proceed to indiscriminate means of violence. For instance, the decision made by Winston Churchill and Arthur 'Bomber' Harris, the commander-in-chief of the Bomber Command during WWII, to proceed to 'area bombings' of German cities – a euphemism for the large-scale indiscriminate bombings of cities with the sole aim of terrorising the population and crushing morale – was clearly a case of state terror[13].

---

must make an intense effort, appoint a troika [a team of three] of dictators, immediately proclaim mass terror, shoot and deport hundreds of prostitutes who intoxicate soldiers, former officers, etc. . . . You must act fast: mass perquisitions. Shooting for keeping of arms. Mass deportations of mensheviks and unreliables. Change the guard at the warehouses, appoint reliable ones. Yours Lenin'. Trotsky shared the same enthusiasm for this tactic. He wrote that 'terror can be very efficient against a reactionary class which does not want to leave the scene of operations. Intimidation is a powerful weapon of policy, both internationally and internally'.

13 In this regard, we can quote Churchill's famous 8th July 1940 letter to Lord Beaverbrook, the Minister of Aircraft Production, in which he wrote the following: 'In the fierce light of the present emergency the fighter is the need, and the output of fighters must be the prime consideration till we have broken the enemy's attack. But when I look round to see how we can win the war, I see that there is only one sure path. We have no Continental army which can defeat the German military power. The blockade is broken and Hitler has Asia and probably Africa to draw from. Should he be repulsed here or not try invasion, he will recoil eastward, and we have nothing to stop him. But there is one thing that will bring him back and bring him down, and that is an absolutely devastating, exterminating attack by very heavy bombers from this country upon the Nazi homeland. We must be able to overwhelm them by this means, without which I do not see a way through. We cannot accept any lower aim than air mastery. When can it be obtained?'. Churchill knew full well the immoral nature of these bombings, which was emphasised in his 28th March 1945 letter to General Ismay following the bombing of Dresden in the closing weeks of the war. He wrote: 'It seems to me that the moment has come when the question of bombing of German cities simply for the sake of increasing the terror, though under other pretexts, should be reviewed. Otherwise we shall come into control of an utterly ruined land. We shall not, for instance, be able to get housing materials out of Germany for our own needs because some temporary provision would have to be made for the Germans themselves. The destruction of Dresden remains a serious query against the conduct of Allied bombing. I am of the opinion that military objectives must henceforth be more strictly studied in our own interest rather than that of the enemy. The Foreign Secretary has spoken to me on this subject, and I feel the need for more pre-

Therefore, defining terrorism as a tool used solely by non-state actors is misleading. History has shown us that the indiscriminate use of violence has also been a feature of state actors that has in fact led to a significantly greater number of deaths than when this method has been used by non-state actors. It is also a mistake to restrict the logic of terrorism to non-democratic states as democracies have also resorted to this type of violence in the past. This is why it is semantically and historically more accurate to define terrorism as the deliberate use of violence against innocent people for the purpose of forcing an entitiy to pursue a specific course of action[14]. Such a definition not only allows us to make a clearer distinction between the forms of violence that can be used by states, but also by non-state actors. It also allows to distinguish between non-state actors that ought to be labelled as "guerrilla groups" whose methods are based, similarly to state actors waging war, on a distinction between combatants and non-combatants from those that are refusing to do so by resorting to indiscriminate forms of lethal measures. When it is the case, such individuals or groups ought to be labelled as terrorist. Although we cannot ignore that this latter form of violence has been the privileged method of some individuals and organizations in the past, it has nonetheless remained a marginal pattern throughout history. In fact, when thinking about the history of non-state political violence, guerrilla appears to have been the norm while terrorism is rather the form of violence that has marked the last 20 years. In this sense, the opposition between "old" and "new" forms of terrorism should rather be understood as a shift from guerrilla tactics to the indiscriminate use of violence because, in itself, the essence of terrorism has remained the same over time.

––––––––––

cise concentration upon military objectives, such as oil and communications behind the immediate battle-zone, rather than on mere acts of terror and wanton destruction, however impressive'.

14 I wish to emphasise the importance of the expression *deliberate use of violence against innocent people* in the sense that the violence these people are victim of is not simply collateral damage. On the contrary, violence against them is the primary desire of terrorist groups, which is what defines terrorism and makes it morally wrong. This is a point of view shared by authors like Christopher J. Finlay, Michael Walzer, Tony Coady, Igor Primoratz, and David Rosenbaum. Finlay restricts the use of the term "terrorist" to individuals or groups who are resorting to deliberate violence against individuals who ought to be thought as immune from any form of violence during an armed conflict (2015, pp. 5–6). For his part, Walzer defines terrorism in the following way: "Randomness is the crucial feature of terrorist activity. If one wishes fear to spread and intensify over time, it is not desirable to kill specific people identified in some particular way with a regime, a party, or a policy. Death must come by chance to individual Frenchmen, or Germans, to Irish Protestants, or Jews, simply because they are Frenchmen or Germans, Protestants or Jews, until they feel themselves fatally exposed and demand that their governments negotiate for their safety (2004, p. 197).

The fact of the matter is that if trying to cause the death of hundreds or thousands of innocent civilians may have been the *modus operandi* of Al-Qaeda and the Islamic State, this method of warfare cannot be extrapolated to all other organizations that have been labelled 'terrorist' in the past. If bin Laden's actions were morally repugnant as they were synonymous with the indiscriminate mass murder of men, women and children, this was not the case with many other organizations in history, which has major implications for legitimate counter-terrorism measures. In fact, contrary to what we may think, most – but not all – of what has been referred to as "the old terrorist organizations" were always keen to base their operations on a discrimination between those they considered legitimate targets and innocent civilians (Neumann, 2009). For instance, the Jewish Sicarii, who are known today as one of the first groups to use terror for the sake of achieving a political objective in an organised way, took great pains to determine who was to be targeted. The immediate cause of their revolt against the Romans was the census implemented in the early years of the Judeo-Christian tradition, which the Jews viewed as a humiliation, a sign of their subordination to a foreign empire. They were able to canalise that frustration into violent actions solely against the Roman 'invaders' and their Jewish collaborators, who were identified as traitors, which took the form of an urban guerrilla. According to the Romano-Jewish historian Titus Flavius Josephus, the Zealots' main tactic was the public assassination of political and religious public figures to create a general feeling of insecurity in Judea[15].

The case of the Assassins shares many similarities with the one of the Sicarii. This sect associated with Isma'ilism also resorted to the public assassinations of public figures who were opposed to their mission. Their most famous murder is, without doubt, the one of Nizam al-Mulk, the vizier of the Seljuk Empire in 1092,

---

15 As Josephus wrote: 'But while the country was thus cleared of these pests, a new species of banditti was springing up in Jerusalem, the so-called sicarii, who committed murders in broad daylight in the heart of the city. The festivals were their special seasons, when they would mingle with the crowd, carrying short daggers concealed under their clothing, with which they stabbed their enemies. Then, when they fell, the murderers joined in the cries of indignation and, through this plausible behavior, were never discovered. The first to be assassinated by them was Jonathan the high-priest; after his death there were numerous daily murders. The panic created was more alarming than the calamity itself: every one, as on the battlefield, hourly expecting death. Men kept watch at a distance on their enemies and would not trust even their friends when they approached. Yet, even while their suspicions were aroused and they were on their guard, they fell; so swift were the conspirators and so crafty in eluding detection' (Jewish War, Book. 2, par. 252–258).

which remains one of the first great terrorist attacks[16]. Subsequent murders included the one of Conrad of Montferrat, the King of Jerusalem, in 1192.

Terrorising policymakers and other public figures as a way to force societies to pursue a specific course of action was the main tool used by 19th century anarchists. In this vein, we can think of the murders of French President Marie François Sadi Carnot in 1894, of Spanish Prime Minister Antonio Cánovas del Castillo in 1897, of the Empress Elizabeth 'Sissi' in 1898, of Italian King Umberto I in 1900, the attempts against leading French politicians Leon Gambetta and Jules Ferry in 1881 and 1884, respectively, or the one committed in 1893 when Auguste Vaillant threw a bomb inside the French National Assembly[17]. The same logic was picked up in the US, where anarchists who opposed the capitalist system killed or attempted to kill individuals who personified in their eyes the oppressive bourgeoisie. This was the case in 1892 when Henry Clay Frick, the director of the Carnegie Steel Company, was the victim of an assassination attempt, or in 1901 when American William McKinley died after being shot by anarchist Leon Czolgosz.

The targeting of these individuals was not personal hatred or a desire for private revenge, but rather their perceived exemplification of the state being fought. This is why some groups – like *La Mano Negra* in Spain – extended their vendetta to local notables from 1882 until 1886. It is true that some groups did not hes-

---

16 As reported by Ata-Malek Juvayni in his *History of the World Conqueror:* 'Now at the time when Hasan first rose in rebellion Nizam-al-Mulk Hasan b. 'Ali b. Ishaq of Tus (may God have mercy on him!) was Malik-Shah's vizier. With his penetrating glance he beheld on the features of the actions wrought by Hasan-i-Sabbah and his followers the signs of troubles in Islam and perceived therein the indications of disturbances; and he strove his hardest to excise the pus of the Sabbahian rebellion and exerted every effort in equipping and dispatching troops to suppress and subdue them. Hasan-i-Sabbah spread the snare of artifices in order at the first opportunity to catch some splendid game, such as Nizam-al-Mulk, in the net of destruction and increase thereby his own reputation. With the juggling of deceit and the trickery of falsehood, with absurd preparations and spurious deceptions, he laid the basis of the fida'is. A person called Bu-Tahir, Arrani by name and by origin, was afflicted "with the loss both of this world and of the next", and in his misguided striving after bliss in the world to come on the night of Friday the 12th, 485 of Ramazan, [6th of October, 1092] he went up to Nizam-al-Mulk's litter at a stage called Sahna in the region of Nihavand. Nizam-al-Mulk, having broken the fast, was being borne in the litter from the Sultan's audience-place to the tent of his harem. Bu-Tahir who was disguised as a Sufi, stabbed him with a dagger and by that blow Nizam-al-Mulk was martyred. He was the first person to be killed by the fida'is. (...) And from then onwards he used to cause the emirs, commanders and notables to be assassinated by his fida'is one after the other. On this account the local rulers (afbab-i-atraf), near and far, were exposed to danger and would fall into the whirl pool of destruction' (1958, pp. 676 – 678).

17 Which only wounded one Member of Parliament.

itate to resort to terrorist acts that were not as discriminate as the ones described before. This was famously the case in France between 1892 and 1894 with François-Claudius Koenigstein, aka Ravachol, whose actions led to civilians being wounded. In his defence, this was not his intention; he was, on the contrary, aiming at state officials who had been directly involved in decisions he felt were unfair. The lack of discrimination of his attacks rather resulted from his methods, namely the use of dynamite, and he openly regretted hurting civilians after the fact (Chaliand & Blin, 2015, pp. 170 – 172). This method was severely criticised by other anarchists for not respecting the discrimination between 'legitimate targets' and innocent people. That was, for instance, the case with the French anarchist Émile Henry, who criticised Ravachol's actions by saying that 'the true anarchist ought to kill his enemies; but he is not doing so by blowing up houses where women, children, members of the working class and servants are living' (Chaliand & Blin, 2015, p. 172). This is what prompted him to detonate a bomb in front of the office of a mining company in November 1892. During his trial two years later, he explained that prior to the execution of his plan he considered the risk of innocent victims being harmed, but quickly came to the conclusion that since the building was only occupied by members of the bourgeoisie only legitimate targets would be victims of the attack. Henry nonetheless ended up following Ravachol's path when he detonated a bomb in February 1894 in a café next to Saint-Lazare train station in Paris. He said in his final declaration that because the authorities resorted to mass and indiscriminate measures against anyone suspected of anarchist sympathies following Auguste Vaillant's attack, the anarchists were also justified to strike indiscriminately (Gazette des Tribunaux, April 29, 1894, p. 419).

Russian anarchists who fought the Tsar's despotism also took great care not to blindly commit murder; they only targeted those who were the incarnation of what they were fighting. Starting from 1866, which was the first assassination attempt against Tsar Alexander II, Russian anarchists also targeted ministers (like Nikolay Bogolepov, Minister of National Enlightenment in 1901 or Vyacheslav von Plehve, Minister of the Interior in 1904). Moreover, when these terrorists were confronted by the decision to kill their target at the cost of harming innocent individuals, they decided to show restraint. This was the case in 1905, when anarchists killed Grand Duke Sergei. Two days before the attack, the one who had been chosen to pursue the mission decided not to throw the bomb when he saw that the aristocrat was accompanied by his wife and nephews, a decision that was later approved by his co-conspirators[18]. The same can be said about

---

18 Two months before the attempt that finally costed the life of Tsar Alexander II, he was tar-

Gavrilo Princip who, having killed Franz Ferdinand, showed remorse for murdering the Archduke's wife, who was not the intended target[19]. As argued by Chaliand and Blin, whether or not we agree with their actions, we must recognize that most of these terrorist groups followed ethics of warfare very similar to the principle of distinction found in the *jus in bello* category that states have promised to uphold (2015, pp. 228).

Alongside the political motivation that drives their attacks, it is this willingness to discriminate between those who are considered legitimate targets from those who are not that sets apart guerrilla fighters from criminals. It is from this perspective that Brazilian revolutionary Carlos Marighela wrote in his famous *Minimanual for the Urban Guerrilla* (1969) that:

> The urban guerrilla, however, differs radically from the criminal. The criminal benefits personally from his actions, and attacks indiscriminately without distinguishing between the exploiters and the exploited, which is why there are so many ordinary people among his victims. The urban guerrilla follows a political goal, and only attacks the government, the big businesses and the foreign imperialists (Marighela, 1969).

Even though terrorist actions are as politically motivated – the establishment of a caliphate or the 'liberation' of a holy land from foreign presence are political goals – as those of guerrilla fighters, the indiscriminate nature of the attacks of the former sets them radically apart from those of the latter. This is why the destruction of Pan Am flight 103 over Lockerbie in 1988 or the mass bombings of London in 1940 by the German *Luftwaffe* and then with V1 and V2 rockets are inherently different from shock attacks by the Irish Republican Army (IRA) against British soldiers. These two forms of violence are different because the first one assumes the civilian hinterland is a part of the battlefield, while the second one refuses to do so (Gross, 2006, p. 1). As such, while guerrilla tactics are

---

geted by members of the *Narodnaya Volya* organization who mistakenly blew up a train in which they thought he was sitting in. As it was reported: "[one of the co-conspirators] remembered how Sofya Lvovna came to Moscow's safe house on November 19 late in the evening. Perovskaya, with a sob, threw herself on her neck, intermittently explaining, "that they had made a mistake, that they blew up the wrong train and that innocent victims were likely to have perished...". Then Sofya Lvovna fell silent and sat in silence for a long time until A.D. Mikhailov arrived. He had brought with him newspapers indicating that their action did not killed anyone. Upon receiving that news, the girls experienced an unforgettable relief and gradually began to come to life" [translation from Russian]. https://istoriki.su/istoricheskie-temy/rossiyskaya_im periya_v_xix_veke/620-dinamit-i-narodnaya-volya.html

19 He said: 'Yes, I am sorry that I killed the wife of Franz Ferdinand. Next to him was sitting the Governor of Bosnia and Herzegovina (Oskar Potiorek), so I shot at him. But it turned out that I killed her. For the rest, I don't regret anything'.

performed in accordance with the customary laws of warfare, terrorist acts are not. This is why the general assumption that guerrilla warfare and terrorism refer more or less to the same reality is inaccurate.

1.2 Understanding the Terrorist Shift

This ethical principle was followed by most 20th century terrorist groups. In this regard, we can speak of the early days of the IRA, whose members purposefully targeted active representatives of the British state who were considered to be combatants (a more thorough discussion about this status will follow in the next section). For instance, on the eve of Easter 1920, the IRA launched more than 300 attacks against police stations and, a few months later, killed 11 individuals working for the British secret services. Under the leadership of Michael Collins, members of the IRA were even sent to Liverpool, where they killed two members of a counter-terrorism unit. In Italy, the Red Brigades followed the same pattern by targeting individuals who incarnated in their eyes the symbols of capitalism. To shake up the working class, its members went on strike against multinational corporations by kidnapping the CEO of Fiat in 1972, a director of Alfa Romeo, another one from Fiat a year later and, more famously, Aldo Moro, the Italian prime minister in 1978.

It is difficult to establish precisely when this ethical code for distinguishing legitimate targets from civilians was abandoned by individuals and terrorist organizations. However, by the mid-1990s, the shift had definitely occurred, with organizations, like the Armed Islamic Group or the Japanese sect Aum Shinrikyo, staging attacks with the clear purpose of killing as many civilians as possible. In the case of the former organization, it famously hijacked Air France Flight 8969 in December 1994 with the purpose of crashing it in Paris and also organized a series of terrorist attacks in the French capital from July until October 1995, resulting in eight deaths and the wounding of 200. For its part, on 20th March 1995, members of the latter organization dispersed sarin gas in the Tokyo subway – one of the busiest commuter transport systems in the world during rush hour – that killed 12 people and injured over 1,000. Of course, the 9/11 attacks that led to the killing of almost 3,000 people remain, in most people's minds, the quintessential example of an indiscriminate terrorist attack. Therefore, we must be careful in the way we assess terrorism by distinguishing between groups that make a distinction between legitimate and illegitimate combatants, a point that had already been made decades ago by Walter Laqueur, who argued that:

Many terrorist groups have been quite indiscriminate in the choice of their victims, for they assume that the slaughter of innocents would sow panic, give them publicity and help to destabilize the state and society. However, elsewhere terrorist operations have been quite selective. It can hardly be argued that President Sadat, the Pope, Aldo Moro or Indira Gandhi were arbitrary targets. Therefore, the argument that terrorist violence is by nature random, and that innocence is the quintessential condition for the choice of victims, cannot be accepted as a general proposition; this would imply that there is a conscious selection process on the part of the terrorist, that they give immunity to the "guilty" and choose only the innocents (Laqueur, 1987, pp. 143–144).

It is even more difficult to explain why individuals and groups abandoned guerrilla warfare and chose terrorism instead. However, I would suggest that it is primarily a matter of strategy which plays – contrary to the popular assumption that groups resorting to these forms of violence are irrational – a fundamental role in whether or not groups will resort to guerrilla or terrorist tactics. There is obviously a different tactic at play here, even though both sorts of actions have the same roots, namely the fact that guerrilla and terrorist groups are facing an overwhelmingly stronger enemy who they cannot defeat in a conventional battle by concentrating all their forces at a specific time and place in order to win a decisive battle. Similarly, to the tactics behind non-violent resistance (Holmes, 1989, pp. 260–294), the resort to elusive methods of fighting are for them the only solution they can consider as a way to cancel out their enemy's superiority and by forcing them to change their course of action as the resistance from the guerrilla/terrorist group makes it no longer profitable by draining its administrative and military manpower. This method, which has been summarized as such in the *Handbook for Volunteers of the Irish Republican Army Notes on Guerrilla Warfare* (also known as the 'Green Book' (1956), also applies to terrorist organizations. It states:

A small nation fighting for freedom can only hope to defeat an oppressor or occupying power by means of guerrilla warfare. The enemy's superiority in manpower, resources, materials, and everything else that goes into the waging of successful war can only be overcome by the correct application of guerrilla methods (Chapter 2). The guerrilla will not fight the enemy in a long battle where reserves would overwhelm him: he strikes only when he can win. And he avoids superior forces. When the enemy advances, he withdraws. When the enemy rests, he hits him. He attacks when the enemy is exhausted. And when the enemy counterattacks, the guerrilla flees' (Chapter 4). 'The regular soldier is no match for the trained guerrilla in attack. Because the guerrilla holds the initiative, strikes when he is ready, uses shock action and surprise to attain his ends, then breaks contact and withdraws (Chapter 8).

A similar tactic was also advocated by Carlos Marighela, whose theories on urban guerrilla warfare resemble those of the IRA. Surprise attacks with lightning results, unconventional methods and draining of the enemy's will to pursue combat are the chief words found in his manual. As a consequence, guerrilla and terrorist groups will usually be organized in a similar fashion, that is through decentralized cells independent from one another and composed of very few individuals (15 to 25 men). These groups usually decide on their own their targets, and how they will carry out shock attacks that are strategically simple, without receiving orders from a central command or any sort of outside help. This was the case with the IRA's Flying Columns and with the terrorists who planned and executed the 2004 and 2005 attacks in Madrid and London. Their goal of such groups is to shock the enemy by perpetrating what appears at first to be an impossible task. Due to the risk of information leaks, only in exceptional circumstances will these groups be called on by this central authority to carry out certain operations, as was the case with the 9/11 attacks by al-Qaeda operatives, as well as on August 27, 1979, when the IRA killed in a coordinated attack at two separate locations both Lord Mountbatten and 18 British soldiers in an ambush now known as the Warrenpoint attack.

Guerrilla groups tend to think of themselves as the spearhead of the cause they are fighting for and as educators of the people by exposing the lies of the enemy they are fighting. This means that gaining and maintaining people's support is essential to their success – a task that is made very difficult by the fact that it is ultimately the population that will most likely bear the weight of the enemy's retaliation measures. In order to succeed, guerrilla groups will therefore need to have strong ties with the civilian population to win its trust, by converting the people to their cause and by supporting and helping them when they are the victims of the enemy's actions. For these groups, victory is not thought possible without that support[20]. As Ho Chi Minh once said, in order to succeed, a guerrilla group is like a fish that needs water to survive and, in this case, water refers to popular support. Alongside the moral reasons that justify in their mind the necessity to discriminate between legitimate targets and non-combatants/civilians, the requirement to maintain the support of the civilian population is another reason that explains the selectiveness of their targets and attacks. Moreover, through their tactics of surprise shock attacks, guerrilla groups are also able to gain civilian support because their actions will provoke an imprecise and indiscriminate response from the state forces that will further reinforce the propaganda concerning the immorality of the state they are oppos-

---

20 This necessity had been identified in the earliest known work on guerrilla warfare by Sextus Julius Frontinus (who died in 103 AD) that the success of guerrilla groups depended on the support of the local population (see Segdwick, 2004, p. 801).

ing. Bloody Sunday of 1920, that led to the death of 30 people – some killed in a stadium during a Gaelic football match – is a good example, because this infamous day was the response to the assassination of 15 British undercover agents by Michael Collins's men the day before. As a result, this act of senseless and indiscriminate retaliation simply increased support for the IRA and radicalized public opinion (Hopkinson, 2004, p. 91; English, 2003, p. 17). As empirical analysis has shown, these groups who resort to either guerrilla or terrorist violence are especially good at selecting such means of actions that will simultaneously limit the capacity of the state to retaliate precisely and effectively against their members and increase the likelihood that the state's retaliation will harm civilians (Carter, 2016). Again, contrary to what may be considered general opinion, these groups are not acting in an irrational manner.

This calculated strategy pursued by these groups may provide a partial explanation of why some of them are shifting from guerrilla tactics to indiscriminate violence. Groups that are operating on the territory they wish to free from the oppression they are fighting must be able to maintain the support of the people by being careful in avoiding harming any civilians during their shock attack. This may explain why guerrilla tactics rather than terrorist acts have largely been used in Northern Ireland, in Spain by the ETA, or in the 1960s in Quebec with the *Front de libération du Québec*. On the other hand, groups operating on foreign soil and bringing combat to the heart of their enemy's territory may not have this sensitivity, since those who might be killed are not considered as "their people" and, accordingly, do not really care if the resorting to terrorism against them results in their loss of support, which they do not require in the first place. As long as such actions are welcomed by their own people, resorting to indiscriminate violence may appear in their eyes to be the best strategy. We can think in this regard to the 1995 bombings in France, carried out by the Armed Islamic Group of Algeria, who planted handmade explosive devices in the Parisian metropolitan train system and which resulted in the deaths of 8 people and 157 people injured; those of radical Palestinian organizations that resorted to suicide attacks on buses and in restaurants on Israeli territory following the second intifada; or of the 9/11 attacks by al-Qaeda.

Furthermore, resorting to such actions can also be the result of a specific dynamic within the society from which the violent activists are coming from. For instance, terrorist attacks greatly increased in Israel following the failure of the Oslo and the Camp David II Agreements, the Palestinians' growing disillusionment with Yasser Arafat's leadership, the provocative visit of Ariel Sharon to the Temple Mount in Jerusalem, and the casualties inflicted on Palestinian civilians by the Israeli armed forces, all favored a surge in support for terrorist actions on the part of Palestinians (2/3 of them showed support for suicide opera-tions) (Bloom, 2004, p. 65).

While some groups may realize that terrorism is strategically counter-pro-ductive, since it tends to alienate the population from the cause they are fighting for and will generate a backlash analogous to when state forces retaliate by harming civilians, as was the case in the aforementioned case of the 1920 Bloody Sunday, others may come to the conclusion that a provocation strategy aimed at targeting their enemy in an indiscriminate fashion will ultimately be favorable to their cause. This is, for instance, the strategy used by the Armenians at the end of the 19th century when they chose to attack Ottoman civilians. However, their strategy did not lead to the expected outcome. As Laqueur wrote:

> The proponents of immediate action prevailed, and since they could not possibly hope to overthrow the government, their strategy had to be based on provocation. They assumed, in all probability, that their attacks on the Turks would provoke savage retaliation, and that as a result the Armenian population would be radicalized; more decisive yet, the Western powers, appalled by the massacres, would intervene on their behalf as they did for the Bulgarians two decades earlier. Lastly, they seem to have hoped that their example would lead to risings among other nationalities in the Ottoman empire, as well as perhaps inspiring disaffected Turks. Their most spectacular action was the seizure of the Ottoman Bank in Constantinople in August 1896. But the results were disastrous: a three-day massacre followed in which thousands of Armenians were killed [and] Europe showed "murderous indifference (Laqueur, 2001, p. 44).

In the same perspective, we can say that the desire to provoke a strong retaliation from the United States in order to generate additional support for its cause was also a part of al-Qaeda's plan in September 2001 (which operated alongside an-other motivation based on an idea of "holy terror" that will be discussed in the next chapter). Indeed, as it was argued by Mark Sedgwick:

> The potential supporters of al-Qaeda's desired insurrection (against regimes such as the Saudi one and against the United States) are, of course, the world's Muslims, or at least the world's Arabs. To the world's Arabs, 9/11 certainly demonstrated the vulnerability of the United States in the most dramatic fashion. By its responses in Afghanistan and – especially – in Iraq, the United States then alienated al-Qaeda's target audience from the United States (…). There were many justifications for those responses, but their impact on the Arab world has been reminiscent of the impact of British policy on Ireland in the aftermath of the Easter Rising. To the average Arab, the toppling of the Taliban in Afghanistan appeared as an act of revenge on the people of Afghanistan, and the invasion of Iraq appeared as an unprovoked attack on a long-suffering people whose only crime was to be Arab and Muslim (Sedgwick, 2004, p. 803).

Such a strategy is, however, a double-edge sword, since it may not lead the vic-tims of such a terrorist attack to resort themselves to indiscriminate killings. If this happens, only the terrorist group will suffer international reprobation, which will lead to long-lasting and permanent reputational harm to its cause. This may explain in part why guerrilla groups that are commonly referred to as "national liberation groups", such as the IRA or the ETA in Spain, have historically shown great care in the design and planning of their attacks. In these instances, these groups are fighting another entity in order to either establish their own country or to impose their ideology. As such, they are aware that the success of their political goal will eventually require the support and recognition of international actors, which is why they cannot afford the risk of being labelled as terrorists, since this would dramatically hinder their capacity to achieve their objective[21]. In some cases, staging a highly publicized attack may also be perceived by some groups as a way to generate international attention for their cause, so that it can create compassion abroad and further increase the pressure on their enemy to satisfy their requests. This strategy can, of course, become even more effective if the entity targeted by the attack resorts to indiscriminate reprisals in the aftermath of the raid. This has been the strategy used by Palestinians when they chose to hijack an El Al flight in Algiers in 1968, or the Armenians in the 20th century who fought for the creation of a pan-Armenian union and the recognition of Turkey's responsibility for the 1915 genocide. As argued by Jenny Raflik (2016, pp. 71–72), the number of attacks against Turkey's allies – namely the United States and Western Europe – decreased at the same time that these countries recognized the existence of this genocide. In return, it is also a key strategic element of entities that are being targeted by this type of political violence to label it as terroristic in order to disqualify guerrilla groups as legitimate actors and to have a broader set of reprisal measures at their disposal. Indeed, by not considering these fighters as legitimate combatants, states are not compelled to treat them as prisoners of war, as was the case with al-Qaeda fighters who were transferred to Guantanamo Bay in the aftermath of 9/11 and were subjected to treatments that shared much in common with torture.

However, these observations are providing a partial explanation to the terrorist option which will also depend upon another important factor, namely the overall goal sought by the group. There is in this regard a difference between groups that need the support of other actors from those that are animated by a transcendental apocalyptic understanding of the world. Groups organized around this latter belief will rather think of their cause as being self-sufficient and of themselves as the instruments at the service of the sacred cause they

---

21 This is why both guerrilla and terrorist organizations usually claim that their fight is one for the noble goals of freedom, democracy, self-determination, and against oppression.

are pursuing. Since everything else is thought to be subordinated to their cause (which includes the judgment of others), their violence is limitless. We can think in this regard to the deadly Tokyo subway sarin gas attack in 1995 by members of the Aum Shinrikyo sect who attacked their own people. In this case, its leader had predicted Armageddon by 1997, following the start of a war between the West and the "Buddhist World" led by Asia during which the forces of evil would destroy themselves and that only the chosen ones would survive (Metraux, 1995). The attack was perceived by its leader – Shoko Asahara – as a way to usher in Armageddon and to offer salvation to those who would die from it. Indeed, "By initiating Armageddon Asahara believed that since Aum's enlightened members were performing the killings, the victims would be purified at their moment of death and would therefore receive the best possible chance of attaining rebirth in one of the higher realms" (Nicholls, 2007, p. 35). This is also the case with Hezbollah, which also sees the death of its enemies as a godly act. When such a mentality predominates, terrorism is more likely to prevail over guerrilla tactics, irrespective of whether or not the group is operating at home or on foreign soil.

However, it must be noted that this role played by religion is by no means a contemporary phenomenon. Indeed, as I have already mentioned, we should not forget that the Thugs' actions – that can be labelled as terrorist considering the indiscriminate nature of their actions – were the result of their willingness to coax Kali and avoid her wrath. From this perspective, we cannot say that apocalyptic religious beliefs are the essence of contemporary terrorism, nor can we say that the indiscriminate targeting of people is a recent phenomenon of the last 20 years. Similarly, the idea that today's terrorist organizations have gone global, contrary to their predecessors, is also an incorrect assessment. As I have argued, the staging of shock attacks outside of groups' national territories is not a recent trend at all. If this has clearly been the *modus operandi* of al-Qaeda or of Islamic State, this was also the strategy used in the 19th and 20th centuries by other groups. In this sense, the global war against terror, in which multiple states are targeted by shock attacks, is not significantly different from when the previously discussed cases of Armenian terrorism – namely through the Armenian Secret Army for the Liberation of Armenia – staged indiscriminate attacks in numerous locations across the world, such as in Turkey (the 1982 Esenboga airport attack) or in France (the 1983 Orly airport attack)[22]. Moreover, it

---

22 It must be noted that this indiscriminate strategy of violence led to a schism within the movement between those in favour of it and those who opposed it and only supported attacks against representatives of the Turkish government.

would also be fallacious to argue that today's terrorism deserves to be labelled "hyperterrorism" because of its unrivalled degree of violence (Heisbourg, 2001). In this regard, we cannot afford the luxury of ignoring the technological means at the disposal of each guerrilla or terrorist group when we are comparing entities that operated at different periods of time. Obviously, having at one's disposal WMD allows to kill a greater number of people with one single attack than with knives or with a vehicle-ramming attack. But, this reality is a technical one and not does not mean that today's groups are more violent and lethal than those from the past. We can in this sense argue that if the latter would have had at their disposal such weapons, they would not have hesitated to use them. Finally, the largely decentralized nature of al-Qaeda or Islamic State, who operate more as a network of loosely connected cells or individuals than as a highly centralized organization, is also not unique in history. As previously mentioned, this organizational strategy was also a feature of the IRA and of Carlos Marighela's manifesto.

Overall, it is therefore difficult to conclude that terrorism has experienced a new and unique shift over the last 20 years, since many of its features were already present before (in some case, centuries earlier), which is why Walzer is wrong when he writes that terrorism (understood as the random targeting of people) emerged as a strategy only after 1945 (2004, p. 198). If we are to accept the distinction between guerrilla warfare and terrorism, we need to agree that this latter form of political violence has not really evolved over time, which makes a discussion about old and new forms of terrorism redundant. What we may be witnessing instead is a convergence of these various aspects that are all coming together, namely an apocalyptic religious understanding of the world that makes winning over the hearts and minds of people by striking only against specific targets a useless strategy which is now facilitated by greater capacities to strike on the enemy's territory and through more violent means. There are, in this sense, no reasons to believe that the world is destined to experience this form of political violence with no hope that it may disappear at some point. Without knowing what the future holds for us, we must keep in mind that a swing of the pendulum may happen and that what we are referring to as contemporary terrorism may disappear almost as quickly as it appeared.

1.3 Justifying the Targets of Guerrilla Warfare

According to what has been discussed, defining a person or group as 'terrorist' implies an objective moral judgment that will prevent making that determination from the subjective perspective of the beholder in a way that always ends up with the cliché that someone's terrorist is somebody else's freedom fighter. Indeed, defining some groups as being terrorist and others as guerrilla simply based upon the political goal they are pursuing is not a proper way of making that distinction. The fact of the matter is that freedom-fighters are not always terrorists. The goal they are pursuing is not directly correlated with their methods of warfare that can be either those akin to guerrilla or to terrorism. Following what I have defended in this chapter, they can be insofar as they are resorting to indiscriminate ways of fighting. In this regard, assessing the respective political goal of groups resorting to violence is not a useful tool in differentiating guerrilla warfare from terrorism.

In this regard, I believe it is better to follow Igor Primoratz's approach by distinguishing groups that refuse to make a distinction between combatants and non-combatants from those that do. If the actions of the former ought to be labelled as terrorist, those of the latter should rather be interpreted under the lens of either political assassinations (Primoratz, 2013)[23] or guerrilla warfare. In other words, the strategy of terror – whether or not it discriminates between those who have lost their immunity and those who have preserved it – plays a big role in our assessment of what is a terrorist group. In fact, we can say that guerrilla and terrorism are interrelated in the sense that the definition of one will inevitably affect how we understand the other. Essentially, the social standing of the victims is what allows us, in my mind, to distinguish between guerrilla and terrorist groups since the indiscriminate targeting of people is semantically associated with the original political meaning of the word *terror* (or *terreur*).

If we can all agree that terrorism as a form of violence can rarely be justified from a moral perspective[24], what about the targeted killing of individuals at the hands of guerrilla groups? And if such actions can be justified, how should we determine which actors have lost their immunity against violence – which can be lethal – from those who ought to preserve their immunity? This is a very tricky question that deserves some attention, although it is not the primary focus of this book. In this regard, even though guerrilla groups are respecting a code of conduct very close to the rules of warfare, it does not mean that their violent actions are automatically justified from a moral perspective. Resorting to lethal actions against other human beings must be grounded on solid moral justifications, which means that some guerrilla groups' assassinations will be justified

---

23 This is also the definition preferred by Stephen Nathanson, who sees the intentional killing of innocent people as the core feature of terrorist groups and what makes them so different from groups resorting to other forms of political violence (2010, pp. 28–29).
24 The only exception would be when a state is facing what Walzer calls a 'supreme emergency'.

while it will not be so for others. How can we make that assessment? Many studies have been published in this regard (Finlay, 2015; Buchanan, 2013; Iser, 2017; Gross, 2015) and have all agreed on the fact that resistance against oppression is justifiable when a state is violating basic human rights. However, this right to resist will take different forms depending on the nature of the oppression. More precisely, the means with which individuals or groups who are victims of oppression can legitimately undertake to defend and protect their natural rights will follow a spectrum that will range from non-violent measures to lethal actions against representatives of the state. This is due to the fact that despite being a catch-all notion, oppression must be assessed based on objective criteria. Otherwise, all forms of oppression – whether it be individuals living under a tyranny, nations denied their right to self-determine freely, the economic domination of the bourgeois claimed by leftist groups, or the cultural hegemony of a so-called imperialist culture – will be considered to have the same meaning and will, as a consequence, justify the same usage of violent and lethal means of actions. Adopting such an expansive view of armed resistance through guerrilla warfare would be problematic as it would justify the resort to lethal actions for members of national minorities denied their right to speak their language, or practice their religion, as well as individuals who live under a genocidal regime that can order at any time their death without warning. Indeed, the right to kill can only be justified in cases of self-defense, that is when individuals' lives are under immediate threat or facing a credible threat against their lives or limbs. While the situation experienced by the Jews during WWII is probably the quintessential example of the first scenario, individuals or groups facing a similar threat by a regime that is displaying the capacity and the intention to fulfil its menace would meet the second scenario, such as the people who opposed Gaddafi in 2011. When these two circumstances are met, the resort to guerrilla tactics is justified. Needless to say, these actual or foreseeable threats must be institutionalized and not the result of violations of individuals' rights by low-level rogue agents of the state. Unfortunately, such actions will almost always happen in societies and the misdemeanor of one police officer or soldier simply cannot alter the nature of a regime that is allowing dissent and its citizens' right to fight oppression – whether it is cultural, religious or economic – through democratic means. In such circumstances, we would not be able to talk about violently oppressive or repressive regimes and, consequently, would not justify the resort to violent actions on the part of those who are the victims of this violation. In other words, contrary to oppressive or repressive regimes that are posing an actual or a predictable threat to individuals' lives or limbs, states that allow for the expression of the frustrations of those who claim to be suffering from forms of oppression have other alternatives than the resort to lethal force. The case of the British rule in Ireland prior to the scission of the island is a good example in this regard. As Christopher Finlay wrote:

> (…) in the late nineteenth and early twentieth century, Britain afforded opportunities to the nationalists of the Irish Parliamentary Party to pursue their political goals through representatives at Westminster who, in turn, promoted their agenda by negotiation with other British parties. This resulted after some decades of agitation in the Third Home Rule Bill, which (unlike its two predecessors) passed successfully through the Houses of Parliament to enter the statute books in September 1914. The Bill did not, by any means, represent the complete fulfilment of the more robust aspirations of some Irish nationalists, but it went some way in their direction and could, perhaps, have provided a basis for further progress in time had the agenda of parliamentary nationalists not been overtaken by separatist proponents of armed force with the Easter Rising in 1916 and the later War of Independence (1919–21). (…) Even assuming that British rule could reasonably be seen as oppressive, the resort to force in these circumstances lacked the necessary prima facie justification since Irish subjects were not by that time subject to Life and Limb Rights violations and those mobilized behind the Irish Parliamentary Party were not subject to Life and Limb Rights violations (Finlay, 2015, p. 83).

However, determining when guerrilla warfare is morally justified as a resort to violence is only one aspect of the equation. The other question that needs to be answered is the identity of those who ought to be targeted by assassinations or shock attacks which can be either assessed based upon an individual perspective or a collective one. Again, this is a tricky question.

In the case of Al-Qaeda and the other aforementioned terrorist groups, it is clear that their very broad understanding of collective guilt does not allow for a distinction between political leaders who are engaging their state in what these organizations believe to be – rightly or wrongly, that question goes beyond the scope of this book – an attack against their most important interests or beliefs, employees of the state without whom these attacks would not be possible, and civilians of all ages and social conditions. These organizations do not view killing these people as murder because they are thought to be either involved in or benefitting from the actions of the state. For instance, a few weeks after 9/11, bin Laden justified the killing of 3,000 people based on the former notion. He said:

> The American people should remember that they pay taxes to their government and that they voted for their president. Their government makes weapons and provides them to Israel, which they use to kill Palestinian Muslims. Given that the American Congress is a committee that represents the people, the fact that it agrees with the actions of the American government proves that America in its entirety is responsible for the atrocities that it is committing against Muslims (Bin Laden, 2005, pp. 140–141).

Such a justification, which gives *carte blanche* to mass indiscriminate murders, is obviously problematic in many regards and is, simply, a piece of ideology with no basis from the standpoint of collective or individual moral responsibility (Miller, 2009, p. 61). First, it is clear that in such terrorist attacks, the chances are very high that individuals who would not satisfy the terrorists' criteria of collective responsibility might be harmed or killed. We can think in this regard of children, but also individuals who may not benefit from what terrorists call an oppressive system or those who are also actively opposed to their country's policy. For instance, when Émile Henry threw his bomb in the Café Terminus in 1894, the place was not entirely filled with bourgeois people; underpaid members of the working class, such as waiters and cooks, were also present. The same can be said when a bomb blows up in a bus or a train in New York, London, or Madrid. People who vote for candidates opposed to what the terrorists see as unlawful policies may also end up being harmed or killed by these attacks. Of course, the terrorists may very well argue that despite their opposition, they nonetheless benefit from such policies, which is enough to make them guilty as accomplices to the crimes of their state. But such an argument is also flawed. Indeed, we need to consider the fact that some of these people are simply unable to emigrate to a country better aligned with their beliefs. Moreover, in the context of authoritarian/totalitarian states, we cannot ignore the fact that being actively opposed to one's state policy always comes with terrible consequences. In this sense, it is worth wondering if it is fair to criticise them for their silence and passivity. Such a conception of collective responsibility would be a major drift away from the current rules of warfare as it would lead us to ignore all distinction during wartime, an idea that was put forward during the 1990 – 91 Gulf War by some members of the US Air Force (Draper, 1998). Indeed, in one briefing, a senior member of this branch of the military said that the Iraqi people could be targeted because they refused to exercise control over their state's policy and did not do anything to undermine their government after it had invaded Kuwait. Such an assessment of collective responsibility would be like assigning liability to all the employees working in a company that produces a life-threatening product, such as a tobacco company or one producing pesticides. This broad and indiscriminate understanding of responsibility is, frankly, dangerous and, if it were to be applied, would simply lead to total barbarianism as it would allow the killing of everyone during wartime.

Morally concerned organizations that refuse to erase the boundaries between legitimate targets and innocent civilians are adopting a less robust version of liability very similar to the one used in the corporate world. More precisely, when companies are accused of negligence for the harm they may be causing, nobody would dare to accuse all their employees of being accomplices or accessories to that damage. Instead, the liability for causing harm always affects some people more than others and the guilt is usually attributed to those who knew or should have known about that negligence. This conceptualisation is usually applied by so-called terrorist organizations that refuse to engage in indiscriminate fighting. Indeed, before striking against an individual, they usually carefully weigh his or her liability for the injustices of the condemned group. It is reasonable to assume that someone may lose his immunity when his direct responsibility has been established with regard to what is seen as a great injustice, great sufferings, or the violation of basic moral rights, such as other people's right to life, not to be physically or psychologically damaged, not to be raped, or to be protected from grievous bodily harm (Primoratz, 2013, p. 17; Shue, 1996; see also Taylor Wilkins, 1992, p. 96). In other words, and similarly to people joining the armed forces, by choosing to defend such a system through their political involvement, they are allowing themselves to be transformed into individuals who have the possibility to implement policies that may hurt and be detrimental to others. What is targeted is, therefore, not their individual personality but rather the institution they are serving, just like in the military where Corporal John Doe is not targeted by the enemy because of who he is as an individual but rather because of his role in the cause he is serving[25]. Those affected by this loss of immunity would, therefore, include political leaders responsible for enacting such policies, those without whom their implementation would not be possible, as well as people who have the power to stop these policies but choose not to do so. This would include tyrants like Muammar Gaddafi, Idi Amin, Saddam Hussein[26], Hitler, or Stalin as well as their henchmen, without whom their crimes would not have been possible, such as Nikolai Yezhov, Lavrentiy Beria, Adolf

---

25 This idea is what Albert Camus had in mind in the dialogue between Dora and Kaliayev in his play *The Just Assassins*. To Dora who says the following: 'One second for you to look at him! Oh! Yanek, you must know, you should be warned. A man is a man. The Grand Duke may have compassionate eyes. You might see him blink, or smile happily. Who knows, he might have a little razor cut. And if he looks at you right then...', Kaliayev replies: 'It's not him I'm killing. It's despotism'.

26 Seumas Miller writes about the former Iraqi dictator: 'Consider Saddam Hussein's refusal to distribute much-needed food and medicine to his own citizens, albeit in the context of UN-sponsored sanctions. Citizens in such states may well be entitled to use deadly force against the government officials in question, notwithstanding the fact that these officials are neither combatants nor the leaders of combatants. Perhaps such use of deadly force, including assassination, is to be regarded as terrorism on the ground that the victims of terrorism are not themselves attackers. If so, then terrorism can be morally justified in some circumstances. However, the civilian victims in this kind of scenario are not innocent; their intentional acts of omission constitute violations of the positive rights of their citizens' (2009, p. 76).

Eichmann, or Reinhard Heydrich (because of his central role in the "Final Solution to the Jewish question", his assassination in 1942 by a team of Czech and
Slovak soldiers was therefore morally justified),[27] but also other low-level enforc-
ers of policies like members of the armed forces, police officers, and civil servants (Miller 2009, pp. 68–75). On the contrary, those who are not in a position
to know about these unjust policies and are not in a position to stop them cannot
be held morally responsible and retain, accordingly, their innocence and right
not to be harmed in retaliation (Caron, 2018b, 2019b). Resorting to deadly
force against the former would be a clear case of self-defence and a way to
put an end to the life-threatening human rights violations that some people
are victim of, while attacking the latter would be morally reprehensible and unjustifiable by any logic. In return, some of the individuals who fall within the
first category have in the past acknowledged that being targeted by these groups
was fair. As was said, for instance, by King Umberto I of Italy after an unsuccessful attack against his life, 'this kind of risk is part of the job' (Primoratz, 2013,
p. 17). Indeed, people who have voluntarily taken on the responsibility of engaging themselves directly in life-threatening rights violations must share the moral
responsibility for their actions, which implies the right of those targeted by these
unjust measures to engage against them in self-defence. Targeting them is, therefore, morally justifiable and should not be considered a terrorist act (Nathanson,
2010, p. 42)[28]. Under such circumstance, Stephen Nathanson writes the following:

---

27 In other words, the loss of one's immunity and becoming a legitimate target would be justified against people who are involved either as a direct perpetrator in the violations of fundamental human rights or a statesman who imposes them. For Seumas Miller, the list of people
affected by this criterion would be: '(…) politicians, or other non-military leaders, who are responsible for the rights violations, or the enforcement thereof, in the sense that in the context
of a chain of command they were the relevant authority that directed that the human rights violations be carried out, or that they be enforced. Such civilians would also include persons who,
while not necessarily part of any formal chain of command, nevertheless, were responsible for
the rights violations (or the enforcement thereof) in that they planned them, and saw to it that
other persons performed the rights violations (or the enforcement thereof)' (2009, p. 69). He also
adds: 'Thus we can make the following claim about collective moral responsibility: if agents are
collectively – naturally or institutionally – responsible for the realization of an outcome, and if
the outcome is morally significant, then – other things being equal – the agents are collectively
morally responsible for that outcome, and can reasonably attract moral praise or blame, and
(possibly) punishment or reward for bringing about the outcome' (pp. 72–73).
28 Alex P. Schmid shares a similar point of view; the target of political violence matters in order
to distinguish between a terrorist act and one that is not. In the case of individuals who have
taken up leadership roles as statesmen or as direct instruments for the implementation of
state policies (e. g. members of the military), they ought to know that they are party to what oth-

An implication of this view is that assassinations are not necessarily instances of terrorism. While assassination is often cited as a terrorist act, the moral innocence definition implies that this is a mistake. Whether an assassination is a terrorist act depends on whether the official who is assassinated had a direct role in creating or continuing the policies being opposed. If an official had no role in these policies, then the assassination would be [an act of] terrorism (Nathanson, 2010, p. 43).

This means that contrary to groups that are viewed under the 'guerrilla' lens, the actions of terrorist organizations are what Alex Schmid rightfully calls the 'peacetime equivalent of war crimes' by striking the unarmed and the innocent not by accident but rather deliberately (1992). Accordingly, fighting these sorts of enemies requires a special strategy that will be discussed in the following chapters. Whether or not we agree with this assessment of liability and whether the potential for lethal consequences associated with deciding on or implementing an unjust policy justifies a total loss of immunity against violence is, of course, a matter that deserves to be debated more thoroughly[29]. Nonetheless, this view has the advantage of providing an answer to the often-mentioned criticism that guerrilla warfare is not comparable to war. After all, the whole tactics of the former tend to differ from those of the latter in a way that makes actions taken during war look more honourable than those associated with targeted assassinations. Indeed, while war is generally understood as a romantic and Clausewitzian idea of opposition between two enemies facing one another, as similarly as a duel (von Clausewitz, 1976, p. 75), proponents of guerrilla warfare have theorised their methods as elusive and based on surprise attacks when the enemy is least expecting them and without wearing any distinctive uniform. In the first case, soldiers appear to act like noble gentlemen, while they appear to be acting like cowards in the latter case. Two things can be said in this regard. Firstly, this is unfortunately a fallacious image of what warfare is about as the idea of fighters facing one another with comparable weapons and the same chances of killing each other is at best a fanciful image. In fact, the military is rather an enterprise dedicated to deceiving the enemy in order to, just like General George Patton once famously said, make sure that only "the other fool will die for his country". In fact, this is one of the main reasons why states are always on the hunt for the development of more sophisticated and precise weapons that

---

ers may see as a conflict and attacks against their lives should not come as a surprise, even when unexpected. On the contrary, civilians who are not directly party to this conflict ought to retain their immunity against harm and death. Groups that nonetheless attack them directly are, therefore, considered terrorist organizations (1992).

29 I am aware that this question is even more debatable when it comes to unfair policies that are not life-threatening, like denying a nation its right to free self-government.

will give them an edge over their foes, even if it means creating a huge asymmetry between them. This view is exemplified by Michael Walzer's idea that there is nothing fundamentally wrong in killing a "naked soldier", that is a combatant unaware that an enemy has the possibility to take his life (Walzer, 2006, pp. 138–139). Secondly, we can rightfully argue that statesmen or soldiers serving in an oppressive or repressive regime actually have a lot in common with guerrilla fighters' methods of warfare. Indeed, the former are also striking against their enemies surreptitiously when they least suspect it. In terms of methods, there are no fundamental differences between snipers hiding on the rooftop of a tall building and who are being ordered to shoot at peaceful protesters in a square, or of René Bousquet who infamously ordered the French gendarme officer under his command as General Secretary of the police to round up in the middle of the night thousands of Jews caught by surprise in their sleep before deporting them to their death in a Polish gas chamber, and those of a guerrilla fighter who is shooting an enemy in the back. In both cases, targeted individuals are unaware of their upcoming fate and, as such, it is hard to sympathise with the perpetrator of a genocide or of mass murder criticism that those trying to kill him are not acting in a "soldier way". By imposing an insidious risk of death on their enemy, they are generating a similar reciprocal right on the part of those targeted by this unpredictable form of violence as an outcome of their right to self-defence.

However, I do believe that the explicit willingness to avoid harming individuals who did not play any direct part in the development or implementation of some collective wrongdoing or injustice deserves to be recognised as a form of ethics and should play a role in response strategies. However, this option is highly questionable when it comes to fighting terrorist groups that refuse to make a distinction between those who are responsible for an injustice and those who are not. Owing to their way of fighting, everybody is, without exception, subject to be killed anytime, anywhere, and without any warning. Because of this *prima facie* moral impermissibility, terrorism must be eliminated, which calls into question states' responsibility to effectively safeguard the lives of their citizens, which remains the most basic and fundamental obligation of liberal states. Unfortunately, as will be discussed in the next chapter, adherence to the Just War Theory approach is of little use and makes it virtually impossible for states to defend themselves and their civilians against these new forms of attacks.

Chapter 2:
The Shortcomings of the Just War Approach Against Terrorism

If we agree that terrorism stands for the indiscriminate targeting of people who have done nothing to justify their loss of immunity against a violent death, we must admit that the terrorist actions we have witnessed in the second half of the 20th century do not constitute a paradigmatic shift when it comes to the political violence of non-state actors. Throughout history, a tension has always existed between this form of violence and that of guerrilla groups. As such, terrorism is by no means an exclusively contemporary phenomenon, as many groups – such as the Thugs – have resorted to the use of indiscriminate harm against people in a way that was similar to the actions of al-Qaeda, Islamic State, or the Japanese sect Aum Shinrikyo. What is now different from the past is the fact that we are living in a world where this form of violence seems to have become not only the new norm, as many historical guerrilla groups have chosen to lay down their weapons in the last 20 years[30], but also the one that is now perceived as an important factor of global instability that can lead to international conflicts. For these reasons, although we can hardly talk about "old" and "new" forms of terrorism, this shift nonetheless constitutes a serious threat that cannot be ignored and that needs to be fought in an effective and moral manner. In light of the options that have been privileged by the Western world in the last two decades, we can say that the way we have fought this form of political violence falls short of these requirements. Indeed, the wars against al-Qaeda and Islamic state have not only left in shreds an entire region, which remains to this day highly unstable and a fertile ground for the rebirth of old or the creation of new global terrorist organizations but have also led to the death of tens of thousands of innocent people in Afghanistan and in Iraq. Far from being a reflection of the West's lack of interest or care in fighting wars morally, what we have witnessed in the last 20 year is symptomatic of a mentality that has perceived its reaction against this threat within the framework of war, which implies waging full-scale wars or military operations against state entities from where terrorist or-

---

30 For instance, in Northern Ireland, most of the groups have decommissioned their weapons since the 1998 Good Friday Agreement, which has led to an important decline of guerrilla attacks in the region, while the ETA in the Basque Country agreed to a ceasefire in 2010, its disarmament in 2017 and finally its dissolution one year later.

https://doi.org/10.1515/9783110729894-004

ganizations are perceived (rightly or wrongly) to be operating from[31]. As a result, states at war with terror have resorted to using the means and actions that are generally associated with a war logic. For the reasons I have already evoked, this has led to strategic failures and violations of the moral rules of warfare. Accordingly, states that are adopting a war logic in the aftermath of a terrorist attack are resorting to an ineffective and grossly unethical strategy.

The decision to deploy troops on a scale similar to what we have been accustomed to in the case of conventional wars between states is entirely understandable. Indeed, when struck at the heart of one's nation by fearless kamikazes who do not hesitate to use any weapon at their disposal to kill as many people as possible, those being targeted by these attacks will obviously ask for revenge with all the possible means and military strength at their state's disposal. After such attacks, it is unfortunately not the time to call for moderation and for a proportionate reaction against the enemy. People want revenge at all costs and they expect their statesmen to show the most extreme determination in this regard. This was clearly the case after 9/11 in the United States. Waiting to have suffered an attack from a terrorist organization before claiming to have a right of self-defense leaves the door open to such a talionic logic, which is why initiating offensive actions in order to prevent being attacked in the first place is the best course of action that states ought to consider. However, this option is also problematic because terrorist groups are largely irresponsive to non-violent measures short of war or NVATW, since their effectiveness is intimately connected with state entities and not with stateless actors. As a result, the only alternative to war that remains at states' disposal is their capacity to pre-emptively attack these organizations. But this option is also problematic since, as this chapter will show, the *modus operandi* of terrorist organizations makes it very difficult for states to justify resorting to such a measure, which has led to a situation where the line between a pre-emptive attack and a preventive attack (which is illegal under International Law and considered to be an act of war) has become blurred. The primary goal of this chapter is to illustrate the shortcomings of the Just War Theory with regard to contemporary terrorism by focusing on the idea that NVATW have a limited effect against this form of political violence.

---

31 This idea is reinforced by heads of states' speeches in the aftermath of terrorist attacks that were seen as acts of war. This was, for instance, the case with President Bush in his September 20, 2001, address to Congress, in which he said: "On September the 11th, enemies of freedom committed an act of war against our country", or of French President François Hollande following the November 2015 attacks in Paris, in which he said the following: "France is at war. The acts perpetrated Friday evening in Paris and close to the Stade de France are acts of war. (...) they constitute an act of aggression against our country, its values, its youth, its way of life".

2.1 Non-Violent Alternatives to War and Terrorism

Like it has already been stated, even though the Just War Theory focuses on when violence can be justified and on how wars ought to be fought, it nonetheless remains, at its core, a pacifist approach because it only legitimizes violence after the failure of all NVATW[32], such as economic sanctions, diplomacy, naming and shaming, mediation, arms embargoes, and non-violent resistance. The main reason for considering these non-violent options is the extremely high cost of war for civilians despite belligerents' best intentions to obey the moral rules of warfare. In this regard, and as argued by Walzer, there are so many unanticipated consequences associated with warfare that statesmen must consider all possible alternatives before plunging their states into war (2004). Therefore, if more peaceful measures can prove more efficient, they must, of course, be privileged. As discussed by James Pattison, the main criterion that ought to be used to determine the best course of action between war and NVATW is the well-known notion of 'war as the last resort option'. There are many ways to assess this principle, but based upon Pattison's work, we must always assume that since war will most likely always bring more harm than NVATW (p. 223), all other non-military options ought to be considered in view of their actual capacities to effectively force a threat to change its course of action. While there are numerous debates surrounding the effectiveness and morality of the aforementioned non-violent alternatives that will not be discussed in this book, their main problem lies in the fact that these discussions have been based solely on the perspective of strategic actors and whose goals they are pursuing is primarily understood as a rational one and are, accordingly, trying to prevent posing actions that will hinder its realization. The effectiveness of these alternatives is also hugely dependent on the fact that they have been designed to apply against state actors and there are reasons to doubt their effectiveness when it comes to stateless terrorist organizations. If economic sanctions can contribute to weakening a national economy in such a way that a state is forced to change its policy, this is obviously not an option against a stateless terrorist organization whose functioning does not depend on the control of an economic market. If states' policies can be sensitive to public shaming and diplomatic sanctions, the same is not true for terrorist organizations that have made the decision to wage an indiscriminate war against their enemies. Contrary to groups that have chosen to resort to guerrilla warfare and whose strategy depends on the support of the pop-

---

32 Obviously, since it does not exclude violence as a last resort, the Just War Theory's conception of pacifism is not an absolute one.

ulation, terrorist groups do not care if their strategy ends up alienating them from those who are not part of their organization. Furthermore, if states can be affected by sport boycotts (Nixon, 1992), this cannot be the case at all with terrorist organizations, which could not care less about not being able to participate in the FIFA World Cup or the Olympics. If arms embargoes can hinder a state's capacity to arm its population or to threaten international security, they have, on the contrary, very little effect on terrorist organizations that, similar to organized crime, count on the black market to acquire weapons. If non-violent resistance can prove itself successful against a state actor whose occupation of another territory stems from its desire to benefit economically from it (Holmes, 1989), it does not apply to certain non-state actors whose motivations lie elsewhere, or who are unwilling to show any mercy whatsoever to their foes, especially when their actions are driven by an uncompromising opposition between good and evil or an apocalyptic worldview.

As a result, this fanaticism has led these groups – such as the Thugs – to lack any sense of what was optimal from a strategic perspective, since their objective was purely dependent upon an unquestionable transcendental purpose[33]. The same can be said about the Aum Shinrikyo sect or al-Qaeda. Speaking about his 9/11 experience while in Cairo, Mark Sedgwick witnessed among the Muslim community the widely shared impression that the attack was plainly stupid from a strategic perspective[34]. This means that terrorism is more susceptible to becom-

---

[33] As David C. Rapoport wrote: "For the holy terrorist, the primary audience is the deity, and depending on his particular religious conception, it is even conceivable that he does not need or want to have the public witness his deed. The Thugs are our most interesting and instructive case in this respect. They intend their victims to experience terror and to express it visibly for the pleasure of Kali, the Hindu goddess of terror and destruction. Thugs strove to avoid publicity, and although fear of Thugs was widespread, that was the unintended result of their acts. Having no cause that they wanted others to appreciate, they did things that seem incongruous with our conception of how "good" terrorists should behave" (1984, p. 660). Bruce Hoffman adds in the same perspective: "For the religious terrorist, violence first and foremost is a sacramental act or divine duty executed in direct response to some theological demand or imperative. Terrorism assumes a transcendental dimension, and its perpetrators are therefore unconstrained by the political, moral, or practical constraints that seem to affect other terrorists" (1993, p. 2). Audrey Cronin writes for her part that terrorist actions from these groups are solely aimed at "pleas[ing] the perceived commands of a deity" (2002, p. 41).

[34] He wrote: "outside al-Qaeda even the least strategically skilled Arabs realized that an attack on America was not a wise move. I will never forget watching the twin towers burn on an Arabized version of CNN – the CNN transmission had been hastily patched onto Egyptian state television with extempore commentary in Arabic – among a crowd that had gathered around a television in a car showroom in Cairo. The prevalent mood seemed to be one of amazement tinged with fear. 'What would happen now?' was the question on every Cairene's lips over the following

ing an option when the religious dimension tends to take precedence over the political imperatives: a feature that has not been the case with other groups who had an important religious component to their fight, such as the IRA, the Zealots or the Assassins, for instance, who privileged guerrilla tactics. In this perspective, it would be a mistake to presume that the presence of strong religious belief automatically means that terrorism will prevail. Many other factors will play a role in this regard, such as whether or not violence is understood as a sacramental act of divinity, the execution of which is perceived as a categorical imperative. In other words, religion can either serve as a legitimizing force for what a group is fighting for, but which will nonetheless be perceived by its members, first and foremost, as an essentially political struggle that requires a well-thought strategy that will take into account the imperatives of *Realpolitik*. When this logic prevails, guerrilla tactics will most likely be the privileged strategy. On the other hand, when religion is perceived as the sole or predominant aspect of the fight, the risks are higher for seeing groups organized in such a way as to resort to terrorism. The immediate and ultimate objectives pursued by these latter groups are other-worldly and, accordingly, this type of political violence is therefore an act that is not perpetrated for an audience – apart from the god(s) – or as a necessary means for satisfying a broader political end, as would be the case for a guerrilla group which sees assassinations as a way to force their enemy to submit to their demands. For this holy terror, violence is an end in itself and religion provides the justification for all forms of violence, even the most indiscriminate ones.

In this regard, the whole debate as to whether Jihadist terrorist groups, such as ISIS, are "truly religious", in the sense that they are not representative of the true values and ideals of Islam (Blake, 2014; Cole 2015), is irrelevant. Arguing over the true essence and purity of a religious message is a sterile debate. The fact remains that religious beliefs, even the most extreme and marginal views within a holy doctrine, are crucial and are always the core elements providing justifications for the killings of innocent people (Juergensmeyer, 2001). Whether they are disputable beliefs, or selectively reinterpreted views on the part of the group to fit their objectives, theology has been central in ISIS's justifications for sexual enslavement, as well as the way it ought to be organized[35]. As a con-

---

weeks. Some form of retaliation was expected, and feared. It was not expected that 9/11 would in any way benefit the Muslim community" (2004, p. 800).

35 As Rukmini Callimachi wrote in a thorough analysis of rape by ISIS, its leadership not only justified sexual violence against women based upon a narrow and selective reading of the Quran, but also gave this act a spiritual dimension. The article is quoting a former slave who said the following: "'Every time that he came to rape me, he would pray,' said F, a 15-year-

sequence, it makes it impossible to generate any leverage against them in a way that will force them to alter their course of actions and will remain unaffected by anything that is not emanating from their religious beliefs. On the contrary, these measures will in all likelihood simply confirm the meaning and value of their actions. This is why non-violent options, namely non-cooperation with the enemy, is seen as a tactically effective option against a moral enemy, but not against one whose desire to realize the utopia he believes in takes precedence over any other strategic or moral considerations. The identity of one's enemy ought to play a fundamental role in our assessment of non-violent options as philosopher Jan Narveson reminds us:

> (...) it is worthwhile to point out that the general history of the human race certainly offers no support for the supposition that turning the other cheek always produces good effects on the aggressor. Some aggressors, such as the Nazis, were apparently just "egged on" by the "pacifist" attitude of their victims. Some of the S.S. men apparently became curious to see just how much torture the victim would put up with before he began to resist. Furthermore, there is the possibility that, while pacifism might work against some people (one might cite the British, against whom pacifism in India was apparently rather successful – but the British are comparatively nice people), it might fail against others (e. g., the Nazis) (Narveson, 1965, p. 263).

The same can be said with regard to additional non-violent measures that have been advocated by a few authors, such as limiting the vulnerability of our infrastructures that may be targeted by terrorists, by having better intelligence against them and by challenging the conditions that foster this form of violence (Martin, 2002). More precisely, it has been said that industrialized states' dependency on high-technology susceptible to sabotage and to cause mass-scale destruction, such as large dams, big power plants or oil refineries, is akin to leaving one's door unlocked in a neighbourhood where there are robbers operating. As a solution that would deter terrorist attacks, Brian Martin has argued the following: "Instead of large power plants, energy efficiency and small-scale renewable energy sources could be used. Microhydro would reduce vulnerability compared to large dams. Organic farming would be far less vulnerable than monocultures. This sort of analysis can be applied to a range of technologies" (Martin, 2001).

---

old girl who was captured on the shoulder of Mount Sinjar one year ago and was sold to an Iraqi fighter in his 20s. (...) 'He kept telling me this is ibadah,' she said, using a term from Islamic scripture meaning worship. He said that raping me is his prayer to God. I said to him, 'What you're doing to me is wrong, and it will not bring you closer to God.' And he said, 'No, it's allowed. It's halal,' said the teenager, who escaped in April with the help of smugglers after being enslaved for nearly nine months" (2015).

Furthermore, he believes that solving what are considered the root problems of terrorism, namely poverty, inequality, economic exploitation and neo-colonialism, would also contribute to the prevention of terrorist attacks.

What is problematic with this proposal is the fact that it is based on a flawed understanding of contemporary terrorism. Firstly, although in the eyes of a terrorist the destruction of the Hoover Dam may be more tempting than destroying a smaller similar infrastructure on a secondary water stream, terrorist groups are always innovating, transforming or abandoning techniques of political violence in light of the way states implement measures of counter-terrorism. After airports' security was tightened for passengers, the hijacking of planes by terrorists who had been able to sneak guns onto planes was replaced by the bombing of planes – a *modus operandi* that became obsolete when states imposed a tighter examination of checked luggage and their mandatory pairing with boarded passengers – and which was itself then replaced by the simple hijacking of planes with rudimentary weapons before transforming them into flying bombs. It is therefore a form of wishful thinking to believe that limiting states' vulnerabilities will inevitably prevent terrorist organizations from attacking them and their citizens. Moreover, and has just been discussed, there are reasons to have doubts over the apparent correlation between terrorism and social contingencies. As the Aum Shinrikyo case demonstrates, these factors had no impact on its members, who were animated by an apocalyptic vision of the world or an unquestionable opposition between good (their cause, obviously) and evil. Japan was not at the time a repressive political regime plagued by poverty, inequality, exploitation, neo-colonialism or torture. From this perspective, and as it has been argued by Rik Coolsaet (2016) and Roberto Baldoli (2020), the premise that the starting point of terrorism "is fear, due to a situation of conflict, or a grievance", and by the fact that "there are some structural issues, a system, or part of it, that is perceived not to work" which leads "individuals, groups or even states (or better, groups within the state) to feel excluded and threatened, by capitalism or scientists, by a king, an individual, a majority or other religions" (Baldoli, 2020, p. 469) may not be an accurate assessment of why some individuals and groups resort to this type of political violence. What can be called "holy terror" or "sacred terrorism" finds its roots elsewhere than in these factors – an idea which has been echoed by Walter Laqueur, who wrote:

> All investigations have shown that poverty does not cause terrorism and prosperity does not cure it. Most terrorists are not poor and do not come from poor societies. In the Indian subcontinent terrorism has occurred in the most prosperous region (Punjab) and the most egalitarian region (Kashmir), while the poorest regions (such as North Bihar) have been relatively free of terrorism. In Arab countries such as Egypt and Saudi Arabia, and also in North Africa, the terrorists originated not in the poorest and most neglected districts but

hailed from places with concentrations of radical preachers. The backwardness, if there was any, was intellectual and cultural, not economic and social (Laqueur, 2009, p. 181).

A similar logic can be found in groups whose actions are animated by another form of non-religious immanence, namely an ideology. Again, we can say that all groups that have generally been labelled as "terrorist" in the past, such as nationalist, anarchist or those of Marxist obedience, were all animated by a form of profound and sincere ideological beliefs. But similarly to the previously discussed groups whose actions were legitimized by religious beliefs, most of these ideological organizations or individuals were also conscious that their fight had an important political dimension and that, accordingly, they had to show great care in how their violence was framed and against whom it was targeted, in order not to anger the actors they needed to keep on their side if they wished to ultimately achieve victory. These groups will therefore be sensitive to any form of political pressure or non-violent actions that might derail that plan. On the other hand, when the realization of the ideological belief is thought to be an end in itself that must be achieved at all costs, groups motivated by that feeling will not have the same perception over the strategic consequences of their actions. On the contrary, they will be more prone to wage a "total war" against their enemy, embodying the same meaning for those fighting on the side of 'holy terror'. Good example in this regard were the Nazis, whose genocidal actions against the Jews were a costly strategic nonsense (alongside, of course, being profoundly immoral and absolutely inexcusable). Indeed, the Final Solution led to major logistical problems that the Wehrmacht had to face, since the trains required to move the much-needed troops to the Eastern and then the Western fronts, as well as the shortage of food, weapons and ammunition, of which soldiers were deprived, were caused in part by Hitler's anti-Semitic obsession and who used massive amounts of strategic resources – namely trains – to send Jews to extermination camps. As magnificently argued by Yaron Pasher (2015), Hitler's ideological obsession played a significant role in his own downfall. When this mentality prevails, non-violent actions will run the risk of being highly ineffective against these groups. Believing in such an extreme Manichean view of the world also helps explain the motivations of those individuals who take a direct role in indiscriminate killing and who do not shy away from their deed at the last minute. Past examples have indeed shown that a devotional and ascetic devotion to the cause predominates over any other moral or political concerns. This was particularly the case with the 9/11 hijackers who followed until the last minute a strict religious ritual that made them forget that their murderous actions were about to kill innocent people, especially women and children. What is called the "Last Night" document, which was found in Muhammad Atta's luggage, is quite revealing in this regard and shows that the hijackers' actions followed a highly complex sets of religious rituals[36]. The same belief in absolute and categorical ideas is also similar in the case of individuals whose terrorist actions are motivated by non-religious motivations. Members of *Einsatzgruppen* and Nazis who played a direct role in the killing of Jews and other victims of Nazism in death camps showed no emotions at the time of their deeds (sometimes even decades after[37]).

If there are reasons to believe that NVATW against terrorist groups are ineffective, this means that there are two options left at states' disposal: either to react only after having been targeted by an indiscriminate attack or to take pre-emptive measures against them. As previously mentioned, the first option is very problematic since, as has been shown in the aftermath of the 9/11 attacks, retaliation runs the risk of being disproportionate, which will lead to the transfer of the risks to civilians abroad, and of creating a serious political vacuum. However, as the next section will argue, resorting to pre-emptive measures is also problematic because this notion has not been designed to apply to non-state actors.

2.2 The Problems of Pre-emptive Measures Against Terrorism

Following WWII, the UN Charter made a clear and conscious effort to limit the rightful use of force only to exceptional circumstances. The main one was the

---

36 More precisely, "In the 'Last Night' document, the operatives of 9/11 were instructed to concentrate on their intentions, to shave, and to make the ablutions required for a state of ritual purity before leaving for the airport. This is approximately the ritual preparation for a major act of worship such as the pilgrimage. Having thus so-to-speak crossed the threshold into ritual space and time, the operatives were then instructed to make supplications at various points, and 'always be remembering God' [*dhikr*], a standard technique of the Sufis. Much emphasis was put on *sabr* [steadfastness or patience], a major Islamic virtue. Even toward the end, purity of intention was required: 'Do not seek revenge for yourself. Strike for God's sake', admonishes the document, following this admonition with an exemplary tale involving Ali ibn Abi Talib. In short, the whole 9/11 operation was ritualized to the greatest extent possible, and the operatives did not shy away from their task. They kept on target until the last moment" (Sedgwick, 2004, p. 807).

37 Such as Klaus Barbie, the "Butcher of Lyon", who showed no remorse during his trial in 1987 and kept claiming he had done nothing wrong despite being responsible for the deportation of hundreds of Jews during the War.

right of self-defence when sovereignty has been violated by another state[38], which allows states to defend themselves against an aggression and also allows third parties to defend a state whose sovereignty has been violated[39]. In light of this restricted view on allowed violence, 'no other kind of warfare are allowed by international law unless explicitly authorised and endorsed – before-hand – by the United Nations' Security Council (UNSC)' (Orend, 2013, p. 34). However, from a geopolitical and moral perspective, there are fundamental reasons to believe that such reactive measures should be avoided against terrorists. Indeed, when attacked by these groups, the appeal for revenge may take over, as was the case after 9/11, triggering a full-blown reaction resulting in a war that will not only destabilise a whole region but also engender a previously discussed situation where control over the events will be lost and where the uncertainties of war will unfortunately deliver their unwanted fruits. Moreover, as argued by Alex Bellamy, waiting to see one's citizens killed by terrorists before claiming to have a right to react is not only imprudent but also highly immoral and contrary to states' obligations to protect their peoples' natural rights, the most important being the right to life. This is why states ought to have a more proactive stance against the terrorist threat. Unfortunately, the way this can be justified is state-oriented and cannot apply to the menace posed by terrorism.

If the violation of another state's sovereignty is considered to be a *casus belli*, customary international law also allows states to proactively resort to violence when they are facing an immediate or certain threat of aggression. Such a situation is known as a pre-emptive war and finds its roots and principles in the *Caroline* incident of 1837, which involved the sinking of an American steamer op-

---

38 If Article 2(4) of the UN Charter states that 'All members shall refrain in their international relations from the threat or use of force against the territorial integrity or political independence of any state, or in any other manner inconsistent with the Purposes of the United Nations', Article 51 nonetheless allows states to resort to resort to violence in order to defend themselves. It states: 'Nothing in the present Charter shall impair the inherent right of individual or collective self-defense if an armed attack occurs against a Member of the United Nations, until the Security Council has taken measures necessary to maintain international peace and security. Measures taken by Members in the exercise of this right of self-defense shall be immediately reported to the Security Council and shall not in any way affect the authority and responsibility of the Security Council under the present Charter to take at any time such action as it deems necessary in order to maintain or restore international peace and security'.
39 As was the case against Iraq in 1990 – 91 when an international coalition repulsed Saddam Hussein's forces out of Kuwait after the small state had been annexed following a few hours of battle.

erating on the Niagara River by the British forces[40]. However, because firing the first shot can easily be perceived as an act of aggression, it is necessary for states claiming that such an action was triggered by a threat to their sovereignty to prove that it was actually the case. In this regard, the *Caroline* jurisprudence states that the threat in question must have been 'instant, overwhelming, leaving no choice of means and no moment for deliberation'. Walzer has reformulated this idea with three criteria, namely that the enemy displays 'a manifest intent to injure' and 'a degree of active preparation that [makes] that intent a positive danger' in such a way that 'waiting, or doing anything other than fighting, greatly [magnifies] the risk [to the state being targeted by this threat]' (2006, p. 81). For him, the Six-Day War of 1967 constitutes a good example of a pre-emptive attack that met these criteria. Indeed, three weeks before Israel struck the first blow, the UN had announced that its Emergency Force that had served as a buffer between Israel and Egypt in the Sinai since the end of the Suez Canal Crisis was to be withdrawn. Immediately, the Egyptian armed forces reoccupied this territory, while the Egyptian government closed the Gulf of Aqaba and the Strait of Tiran to Israeli boats. At the same time, the Egyptian armed forces were put on maximum alert and mobilised while military alliances were signed with Jordan, Syria, and Iraq. Finally, Gamal Nasser, the Egyptian President, declared on 29th May that in the eventuality of a war against Israel, his 'goal would be nothing less than [its] destruction' (Walzer, 2006 p. 83). Faced with these threats, Israel attacked its enemies on 5th June as it became clear to its government that it was only a matter of days before the country would be under attack.

Justifying this form of anticipated violence is seen by its proponents as a matter of justice against a would-be aggressor who is already guilty by reason of his intent to commit an illegal action that might result – as is unfortunately the case with any war – in the killing of innocent people. As put by J. Warren Smith:

> Lesser evils may be necessary to avoid greater evil. There are two ways in which the greater evil may be identified. First, there is a utilitarian standard. The death of the one brigand is preferable to the deaths of many innocent travelers that will occur if the brigand is not stopped. Second, there is the standard of justice by which each man should receive his due. The killing of the brigand is an evil but it is preferable to allowing the death of an innocent man. The death of the traveler is a greater evil precisely because the traveler is innocent and so does not deserve death. The death of the brigand is a lesser evil because he is al-

---

40 The criteria for a legitimate pre-emptive attack are themselves inspired by the work of Hugo Grotius (1583–1645), for whom anticipatory self-defence ought to be allowed when there is a present danger and a threatening behaviour that is imminent in a point of time.

ready guilty by virtue of his evil intention. The greater evil to be avoided is allowing injustice to be perpetrated (Smith, 2007, p. 146).

According to this conceptualisation, a pre-emptive action ought to be understood strictly as an act of self-defence. Of course, in such cases, it is an anticipated form of self-protection as, contrary to opposing someone or an entity that has already struck the first blow, it is intended to be used against a person or entity that poses an actual threat. The whole question is, of course, to determine when a threat moves from 'potential' to 'real'. After all, if we are to harm individuals who still have not hurt us, morality imposes on us the obligation to prove that our action is justified because our enemy's upcoming attack was certain. As summarised by Suzanne Uniacke, the notion of 'imminence', deemed to have been present in the aforementioned case of the Six-Day War, plays a pivotal role in the assessment of a threat. She writes:

> (...) Striking before one is struck can reasonably be regarded as self-defense only if it approximates closely to an act of retaliation, a return of harm. This is why the imminence of the attack being preempted is significant to the representation of restricted instances of preemption as self-defense. (...) Outside of these conditions, the use of preemptive (preventive) force against a (merely) possible or potential threat is not an act of self-defense. A person acting in self-defense aims to prevent the infliction or imposition of a harm or a wrong; he does so by resisting or repelling an actual, or under certain conditions an imminent threat. In contrast, preventive force aims to prevent a possible or potential threat from becoming an actual threat, by means of disabling a possible or potential threat from become an actual threat, by means of disabling a possible or potential aggressor (Uniacke, 2007, p 80).

Allowing states to defend themselves against such threats bears a lot of similarities with domestic laws. Indeed, while people are not allowed to unjustly attack their co-citizens, they are, however, allowed to defend themselves after they have been attacked or when they feel threatened by a genuine menace with means that would normally constitute a crime. In the case of what would be deemed a pre-emptive attack, people can defend themselves only if the threat is imminent and credible. In this case, simple verbal threats not accompanied by a clear intent and means to fulfil them would not justify the resort to defensive actions.

It is fairly easy to understand why resorting to force outside these peculiar situations can be problematic. Indeed, if force ought to be tolerated in the domestic sphere simply on the basis of someone's fear of eventually being attacked, it may lead to a generalisation of unjustified violence against individuals who may not be motivated by any form of animosity. In the realm of inter-state violence, deadly large-scale wars may end up being waged against political entities under the same excuse, which is problematic as threats are an essential component of the dynamic of the international community. To put it in more simple terms, because of its destabilising effects, war should only be an option when all others have failed or are no longer available. This is why David J. Garren (2019), similar to Hugo Grotius, is right to point out that mere suspicion and fear do not constitute solid grounds to justify resorting to anticipatory self-defence actions that may prove deadly. For the sake of limiting violence as much as possible, self-defence needs to remain a highly restricted option that must be circumscribed by the presence of an ongoing harm against oneself, the certainty that a harm will occur, or the imminence of a harm that can only be repelled with violence[41]. Accordingly, the right to anticipatory self-defence can only exist when a threat against us is 'coming or likely to happen very soon', 'ready to take place' or 'hanging threateningly over one's head' (Lubell, 2015, p. 702).

However, because of the reality of the world order following WWII, this understanding of legitimate violence can work when we are dealing with conflicts between states. However, since the end of the Cold War, with an increasing number of civil wars and those being fought by non-state actors – as in the case of terrorism – this has not necessarily been the case[42]. Thus, the assumptions upon which the Just War Theory was constructed in the post-1945 world are not in tune with certain features of contemporary conflicts. First, it is obvious that contrary to states, the terrorist groups previously discussed are not risk-averse entities against whom the traditional forms of deterrence will work. If, in the past, WMD were an effective way to defend one's sovereignty through the fear of mutually assured destruction, this is no longer the case against these terrorist groups. More precisely, because of their stateless nature and the fact that their

---

41 He gives the following example: 'If you knew that in a week's time you were going to be stricken with an irreversible form of paralysis that would leave you unable to move or speak, and knew that once stricken your enemy was going to take advantage of the opportunity to kill you, would it be permissible to kill him first, in advance of the paralysis, or would you have to wait and take your chances? (...) If you knew to an absolute certainty that you were going to be paralyzed, that it was irreversible, that your adversary was going to kill you and that nothing short of killing him would stop him from doing so, I think that a compelling argument might be made that your right of self-defense would permit such an anticipatory measure. The obligation not to use lethal force first is, after all, a defeasible one and it might well be defeated here, the abundance of certainty (with respect to options and outcomes) compensating for the lack of imminence (with respect to threat)' (Garren, 2019, p. 204).

42 C.A.J. Coady wrote that 'the last quarter of the twentieth century and the beginning of the twenty-first century have seen a dramatic decline in warfare understood as direct state-to-state conflict' (2008, p. 4).

members seek martyrdom through their actions, governments targeted by terrorist groups are in dire need of finding new ways to avoid their attacks (Bush, June 1, 2002). However, this comes with another problem, as the pre-emptive logic is of little use against terrorist groups that are no longer willing to discriminate between combatants and non-combatants. Indeed, contrary to state actors, it is very difficult to effectively prevent a terrorist threat from occurring through the logic of pre-emptive self-defence (Buchanan & Keohane, 2004, p. 3), which makes the war against terror unique. Unless some concrete information about an upcoming terrorist attack becomes available, what is lacking here is the previously mentioned imminence criterion that cannot be assessed with these groups; because of their *modus operandi*, these elusive enemies are able to covertly attack and kill thousands of civilians without any precursory signs. In fact, contrary to state actors planning to violate another state's sovereignty, terrorist groups do not display the mass mobilisation of troops and military equipment alongside their enemy's borders. On the contrary, because of the asymmetrical nature of their fight against great powers, their success relies on the element of surprise, which has been summed up by Noam Lubell in the following way:

> The challenge posed in the context of imminence is that, in effect, we are faced with a threat, for which we cannot positively identify how soon it might happen, where it will originate from, where it will strike, or even who precisely will be behind the attack. (...) the threat of terrorism plays on the fear of the unknown, and raises the question of engaging in self-defense to prevent a possible future attack without knowledge of what it might be. As such, it challenges not so much the interpretation of imminence, but the effect calls into question the very existence of the imminence requirement. (...) the idea of acting to prevent a vague and non-specific threat cannot, therefore, be covered within the concept of imminence (Lubell, 2015, p. 707).

Indeed, sticking with the notion of imminence against this type of threat is a recipe for disaster. Owing to terrorist groups' surprise attacks and their potentially apocalyptic use of WMD, waiting for their threats to become immediate is suicidal (Beres, 1991; Glennon, 2002). This is why Dominika Svarc has argued that:

> The particularly grave threats which could materialize in attack without a reasonable degree of warning and time for defense may be regarded imminent even when the attack is not menacingly near. (...) Applying the narrow temporal standard of imminence in such contemporary reality might deprive a State from an opportunity to effectively repel the attack and protect its population from unimaginable harm. It would go counter to the object and purpose of the right of self-defense which provides States with a self-help mechanism to protect them from an attack when peaceful alternatives would prove inadequate and the multilateral response too tardy (Svarc, 2006, p. 184).

In this sense, it is hard to criticise former President Bush for his assessment of the terrorist threat during his 7th February 2003 State of the Union Address in which he said the following: 'Some have said we must not act until the threat is imminent. Since when have terrorists and tyrants announced their intentions, politely putting us on notice before they strike? If this threat is permitted to fully and suddenly emerge, all actions, all words and all recriminations would come too late'. What needs to be avoided, however, is the radical and inadequate solution of eliminating the distinction between pre-emptive and preventive wars.

Contrary to the R2P logic, which has emerged out of the new types of conflicts that have appeared in the last quarter of a century, the international norm has not been changed in order to allow states to proactively defend themselves against the terrorist threat despite the numerous calls to do so[43]. If states wish to abide by the rules, this basically leaves them with only one option: wait to be under attack before claiming the right to self-defence. This is obviously a highly questionable option from a moral perspective, as states have to paradoxically first sacrifice the lives of their citizens in order to have a right to defend them. On the contrary, we must also admit that terrorist attacks have led to disproportionate reactions on the part of the targets. When states are struck by this type of attack, the natural reaction and the popular pressure will be to seek revenge. In light of what happened in Afghanistan following 9/11, this desire may turn into a full-scale war, an option that has high chances of being ineffective, counter-productive, and leading to a disproportionate number of victims. Almost 20 years after Operation Enduring Freedom, Afghanistan remains a failed state and an unstable powder keg that is still plagued by high levels of corruption and sectarian political violence. However, more importantly, the reason behind the invasion was the incapacity to eliminate the terrorist threat. According to the Charity & Security Network, the Taliban are still behind terrorist attacks in the country and have dramatically grown their number of supporters to more than 65,000 (as of January 2018), while Al-Qaeda has shifted its focus towards the larger Indian subcontinent in order to institute a caliphate and is still planning to conduct attacks against the US and other Western countries. Moreover, we cannot ignore the fact that nearly 50,000 civilians have died in Afghanistan since 2001 either as a result of the war or the breakdown of public infrastructure it induced. In this sense, there is something hugely paradoxical when we consider the reasons why the US intervened in Afghanistan after 9/11. If the main reason was to prevent innocents from being killed by future Al-Qaeda attacks, the decision to wage a full-scale war led to an exponentially greater number of civil-

---

43 For a summary of the proposals made by states in this regard, see Bethlehem 2012.

ian deaths. If the underlying logic of this decision was that American lives are worth more than those of Afghans, then it is, of course, unjustifiable from a moral perspective. The same can be said about Iraq, where the death toll of civilians is at least of 200,000 people either because of the war itself, the failing infrastructure, or killings by the Islamic State (Gollob & O'Hanlon, 2020). Moreover, the 2003 invasion directly led to the creation of the Islamic State, whose affiliated members launched deadly attacks in numerous Western countries. Again, if the intentions of the Bush administration were to prevent terrorist attacks and to protect innocent civilians, they failed miserably.

It is, therefore, easy to understand that between inaction and full-scale wars, to save innocent civilians, there have been numerous calls to revise the legitimate use of violence against terrorist organizations that do not recognise a distinction between combatants and non-combatants. This was the spirit behind President Bush's National Security Strategy following 9/11, which talked about the need to identify and destroy threats before they reached American borders (Bush, June 1; September 14, 2002). If the ill-fated decision to invade Iraq one year later[44] proved problematic since President Bush had in mind the right to wage preventive wars based on false or misleading assumptions, his underlying logic remains valid. If states cannot effectively deflect this threat and protect their citizens through pre-emptive actions, there is a need to correct the situation through proactive actions that will not lead to full-scale preventive wars like the 2003 invasion of Iraq. This implies reviewing the criteria regarding what ought to justify a pre-emptive strike which means replacing the criterion of imminence with another one. This is what the next chapter will attempt to do.

---

44 As mentioned previously, this decision led to political instability, the emergence of a new terrorist threat, and the death of thousands of innocent civilians.

Chapter 3:
Revisiting the Notion of the Pre-emptive Attack

As argued in the preceding chapter, traditional forms of deterrence as well as the logic of the pre-emptive attack are of little use against some contemporary terrorist groups that refuse to discriminate between combatants and non-combatants. This situation is morally problematic as states are no longer able to protect their citizens from harm. The main reason for this failure is the difficulty of foreseeing these threats. As a consequence, since there are reasons to believe that NVATW will prove ineffective, there is a need to rethink the foundations of pre-emptive strikes when a state is threatened by a terrorist organization. Therefore, this begs the question of how states should act when they are facing such a threat. More precisely, how should they distinguish between an unfounded threat and a real one? With regard to the latter point, if these threats are never imminent because of the terrorist organization's *modus operandi*, what criteria should then justify states' reaction? These are fundamental questions that need to be answered if we are to justify forceful actions against these groups and the state entities that may be sponsoring them. As will be discussed in this chapter, this can be achieved by broadening the definition of what constitutes an imminent threat through the prism of its credibility.

3.1 The Need for A New Criterion: A Threat's Credibility

Reassessing the pre-emptive attack logic is highly controversial and does not come without serious concerns since, as said by Deen Chatterjee, the '(...) US war on terror [against Iraq is] an example of what could go wrong with [a more permissive logic of political violence]' (2013, p. 2). The most important of these concerns is the adoption of an overly generous view of what constitutes an imminent threat, which would lead to a legitimisation of wars against entities that are not really a menace, ultimately leading to further destabilisation of the world order by setting off 'a cascading series of "preventive" attacks or interventions' (Bethke Elshtain, 2013, p. 23). However, because of the terrible consequences of terrorist attacks and the inherent difficulties in foreseeing them, there is a need to act before these groups strike, which of course creates an incredible dilemma. In the words of legal theorist George Fletcher, there is a legitimate reason why we ought to avoid an unlawful strategy that prematurely legitimises the resort to force and, contrarily, to come up with an approach that makes retaliation after a terrorist attack has succeeded the sole option (1998, p. 133). This is why

https://doi.org/10.1515/9783110729894-005

we cannot consider all forms of preventive actions under the same lens and why the criticisms usually associated with preventive wars against unfounded enemies cannot apply to those posed by the terrorist organizations described earlier. This renewed approach to violence ought to remain as limited as possible and should be based on an assessment that leaves little doubt about the nature of the threat, which is close to Walzer's understanding of the distinction between non-legitimate and acceptable use of preventive violence that ought to depend on a reasonable perception of the danger of the threat (2006, p. xiv).

Contrary to the right to retaliate after having suffered an armed attack that is considered an act of war, the principle of pre-emption implies the capacity to attack entities before they attack us. In this case, we need to make sure that the decision to attack is morally justified and a genuine matter of self-defence. What needs to be avoided in such a case is a situation where a state uses force against an entity that does not pose a real threat. We, therefore, need to wonder how the reconceptualization of pre-emption might apply against terrorist organizations. Many solutions have been suggested in this regard in the wake of the 2003 intervention in Iraq. For instance, many authors pointed out that the possible collaboration between this rogue state in possession – as it was thought at the time – of WMD and a terrorist organization was clearly a recipe for disaster. As Stephen Strehle put it:

> [Iraq's] relation to terrorism is a matter of grave concern. It provides a special channel to deliver and promote his wicked designs. Bin Laden has called it a "religious duty" for his minions to obtain and use WMD against the infidels, but he knows that his terrorist network needs help. It is only in the movies that Dr No is able to create the facilities to manufacture and deliver WMD. In the real world of terrorism, the capacity to make and utilize these weapons requires the help of a government. Aum Shinrikyo, a Japanese cult, tried to kill thousands of commuters with a potent nerve agent but managed to kill only a dozen after spending somewhere around thirty million dollars. The loss of these lives was tragic but much less than expected and displayed the complexity of operations using these agents. The cult was not able to produce the chemical (sarin) in sufficient purity and resorted to using a most primitive delivery system – carrying it on a train and piercing bags of it with tips of umbrellas. A government working with a terrorist organization would produce a more lethal combination (Strehle, 2004, pp. 77–78).

In the event of such a collaboration between a state and a terrorist organization, the obvious fear is that once the former provides the latter with a WMD, it will already be too late to react as this weapon will vanish until it is used in a surprise attack against thousands of innocent civilians. There is, in this sense, a need to act before the terrorist organization can manage to get its hands on such weapons. For Strehle, no one should be expected to be threatened by such a menace against their lives and 'the mere possibility' that a terrorist organization would be able to have WMD at its disposal justifies its 'immediate elimination' (2004, p. 79). Jean Bethke Elshtain also followed a similar logic in her justification of the intervention in Iraq (even after it took place[45]). For her, because there is a strong possibility that criminal regimes will engage in wrongful actions, the case for intervention becomes stronger. This is, in a nutshell, the logic used at the time by the White House when President Bush ordered the invasion of Iraq. Such a criterion is, unfortunately, highly problematic as it is still pretty much based on a fear that is not or may never be founded in the future. This opens up the risks of interpreting too broadly what constitutes a threat, leading to interventions against unfounded threats, and, as a consequence, the unjustified direct or indirect suffering and deaths of innocent people (Schweller, 1992, pp. 236–237). The lack of evidence that Saddam Hussein possessed WMD and was collaborating with Al-Qaeda as well as the terrible consequences of this invasion should make us wary about considering this criterion a plausible one. In this regard, 'fear' as a criterion should be abandoned as possible grounds to justify the resort to violence against an enemy[46].

In fact, we may even question the desire to eliminate fear in the field of politics. Of course, it is perfectly understandable for people and states to want to live in a peaceful world untampered by the fear of seeing one's security being challenged. However, at the same time, we cannot ignore the fact that it is not only an unachievable political dream, as it is in the nature of other states to always seek to maximise their power and influence over other entities, but that it is also not desirable for societies. Indeed, as argued centuries ago by Augustine, a small dose of fear can be highly beneficial to societies as it allows for the development of well-needed social virtues, namely vigilance and a willingness to protect and defend the institutions that allow people to be free and to enjoy basic human rights[47]. On the contrary, when individuals are no longer fearful, there

---

45 She wrote in 2006 that the intervention was justified (2006, p. 110).

46 Ancient and modern authors were of the opinion that fear has often been the leading cause of wrongs in the past. As Xenophon once wrote in *Anabasis*: 'For I know that there have been cases before now – some of them the result of slander, others of mere suspicion – where men who have become fearful of one another and wished to strike before they were struck, have done irreparable harm to people who were neither intending nor, for that matter, desiring to do anything of the sort to them'. Grotius also shared a similar assessment of how we should be wary of fear as a criterion to justify the resort to violence (*De Jure Belli ac Pacis*, 2.1.5.1). See also Clough and Stiltner (2007, p. 259–260).

47 As a Christian thinker, Augustine primarily saw in fear a capacity to elevate men so they could reach the City of God. As J. Warren Smith wrote in this regard: 'the struggle against fear is part of God's training humanity in the way of true justice. Rather than escaping fear by eliminating its temporal causes, Augustine argues that we must live with our fear so as to place our

is a risk that they will lower their guard and will not be able to foresee a real threat from arising. For Augustine, it is precisely this lack of vigilance caused by their lack of fear (which led to *apatheia*) on the part of Roman citizens that favoured the conflicts between Marius and Sulla or Pompey and Caesar (*The City of God*, I.30). In our liberal democratic era, following the well-known arguments of Benjamin Constant and Alexis de Tocqueville, it is the fear that statesmen might abuse their power and deprive us of our freedom that forces us to remain vigilant to all their comings and goings. Whether we like it or not, there is an undeniable value in fear.

Another solution in this regard would be to argue that it is legitimate for states to attack groups that threaten them and their civilians. Indeed, our right to defend ourselves against people who pose such a menace seems unquestionable. Since, by nature, terrorist groups resort to intimidation against individuals to pressure targeted states to change their policies, and such intimidation usually involves threats to kill or physically harm people, it would appear self-evident that we ought to have the right to strike before they fulfil their promise of destruction. However, this is a nuanced situation. Indeed, in criminal law, threatening to kill or harm someone does not imply that the person who poses the menace ought to lose his immunity against violence. Such an outcome would depend on two important factors, namely the specific nature of the threat and its reasonableness. For instance, a drunk customer in a pub, barely able to stand on his feet and threatening to kill all the other customers if he is not served his drink, would most probably fall short of meeting the threshold for prosecution as no reasonable individual would consider the threat real. The reaction would, however, be different if three lucid, muscled guys armed with guns made the same threat. In this case, the other customers would probably consider the threat credible and genuinely fear for their lives. In this scenario, the public display of weapons pointed in their direction would allow the customers to resort to all possible actions, including deadly force, against the three men, a reaction that would be deemed commensurate with the nature of the threat and justified as a matter of self-defence. On the contrary, using the same level of violence against the stumbling drunk would clearly not be proportionate to the menace he poses, as his worrying words would not be linked in any way with his actual capacity to fulfil his threat. Therefore, having the actual means to achieve one's ambitions is a fundamental variable to consider.

---

hope upon the kingdom to come. There is a sense that living with fear is living with the reality of death and the threat of our judgment by God. Living with such knowledge cultivates humility' (2007, pp. 149–150).

Of course, to respond, one does not need to wait until the attacker's threat becomes a reality, as individuals facing this menace may protect themselves or count on the state's authorities to do so in their name. Indeed, credible threats against people's lives have always justified resorting to potential lethal counter-measures against their perpetrators, irrespective of whether or not they have the willingness to transform their threats into reality. In such a case, the appreciation of the threat and the methods that ought to be used to fight it are in the eyes of the beholder[48]. For instance, let us imagine that a man walking his dog in a park is suddenly threatened by someone holding a handgun, who asks him to choose between his wallet or his life. However, this turns out to be an empty threat as the weapon is either not loaded or loaded with blank rounds. At the exact same time, a police officer on duty witnesses these events and is able to hear the perpetrator threatening the innocent walker. There is no doubt that he would be justified in using all possible means – including lethal ones – against the former individual. In this case, the threat is enough and the police officer is not required to wait until the perpetrator pulls the trigger before acting. If he were to use his service pistol to neutralise the menace and it were to later be discovered that the perpetrator was not actually in a position to harm anyone, it would not make the police officer's decision less legitimate. In appearance, this is exactly the situation with terrorist organizations that make credible threats.

There is, however, a major difference between them and the aforementioned case of the man menacing the individual walking his dog, as in the latter case the threat to someone else's life is *credible, immediate, and imminent* from an empirical perspective. Such a situation is very similar to the cases of countries like Egypt, Syria, and Jordan, that were obviously planning an attack against Israel in 1967. Both of them would fall within the category of a justified pre-emptive attack. However, this is not really the case with terrorist organizations, whose verbal threats are not always perceived as imminent. This is, of course, a problem as words alone do not contribute to transforming a menace into a credible threat that requires immediate action. Consequently, and contrary to the previous case, resorting to pre-emptive counter-measures would not be justified. If we were to justify the resort to war or measures short of war solely based on verbal threats, the world would soon become even more chaotic and violent than it currently is. It is, nonetheless, also a dangerous game to automatically regard all sorts of threats as trivial and to systematically downgrade them simply

---

48 This is also known in criminal law as the 'reasonable person standard', namely that a person is entitled to resort to defensive force insofar as her decision to act as such resulted from a belief any other reasonable person would have had.

to rhetorical hyperbole, especially when these threats are coming from terrorist groups. The crux of the problem with these entities is that there is never any transitional point between their verbal threats and the attack itself. As already stated, it must be noted that their *modus operandi* is unique in the sense that it is basically impossible for states to act pre-emptively when an attack from these groups becomes imminent as their strategy is all about striking at states covertly and taking them by surprise. In the case of the aforementioned example, the terrorist threat will not take the form of an individual approaching him with a handgun and an explicit threat to his life. It will rather take the form of a menace that will never be foreseen by the man until it is too late. More precisely, in this case, the threat will come from a bullet fired from hundreds of yards away by a well-hidden sniper or by a bomb that has been carefully concealed in a garbage bin, set to explode when the man approaches. In this situation, the threat might be real, but it is not imminent to anyone because of its hidden nature.

From this perspective, a warlike rhetoric coming from a state whose decision to transform its words into actions will always be empirically obvious through clear evidence of its willingness to strike, just as is the case with an armed man walking towards you and threatening to kill you if you don't give him your wallet. In the case of states, this threat will take the form of the mobilisation of their armed forces and other preparations that indicate an upcoming attack. If the resort to violence can be justified in such cases, it is not so with threats that cannot be seen. Of course, it is always possible to argue that terrorist groups can be kept under surveillance and that, with good intelligence, it is possible to learn about their upcoming plans. This appears to be in no way different than having a group of criminals under surveillance by police officers. But there is a major difference between these two situations. While in the latter case this surveillance is possible thanks to a well-functioning police force and because the state has the capacity to legislate, monitor what is happening on its national territory, and take effective measures against these groups, this is not necessarily the case with terrorist groups operating from a foreign territory. The states targeted by these groups need to rely on the hypothetical cooperation of other states that may not be keen to share essential information or, even worse, are providing help and support to these groups.

Next, we cannot isolate the secretive nature of terrorist attacks from their potential magnitude. As they are willing to use extreme tactics in order to kill as many civilians as possible, there is an obvious danger in not acting until the threat becomes a reality. This poses a serious moral problem as states facing a terrorist threat can only rely on reactive measures after they have been the victim of an attack according to the rule-based legalist approach of the post-1945 Westphalian paradigm. As was the case with 9/11, this may mean resorting to forceful measures against a terrorist organization, potentially leading to unforeseen consequences such as in Afghanistan, namely an endless war against an invisible threat, the destabilisation of a country, the death of countless innocent civilians, and the incapacity to eradicate the original menace. Moreover, this means that states need to let their citizens die before being allowed to react, which is morally problematic. After all, the duty to protect its citizens is one of the most basic obligations of a state. Indeed, according to our modern tradition, governments are instituted for the purpose of ensuring the protection of their citizens, as has been argued by Thomas Hobbes, John Locke, Montesquieu, and Jean-Jacques Rousseau[49]. In other words, states have a contractual obligation to provide their citizens with an environment that is better than any they could have in a state of nature. This obligation may take various forms. As argued by the European Court of Human Rights, 'This involves a primary duty on the State to secure the right to life by putting in place effective criminal-law provisions to deter the commission of offences against the person, backed up by law-enforcement machinery for the prevention, suppression and punishment of breaches of such provisions" (Kilic v. Turkey, 2000, p. 16). From this perspective, a state that refuses to act against a credible terrorist threat would be like a police officer waiting for a criminal to kill his hostage before taking any forceful measures against the offender. I think most people would see this lack of proactivity as a form of negligence on the part of a state agent, which is detrimental to the political association's duty to protect those in danger[50].

Lastly, the types of weapons terrorist organizations threaten to use against us and their risk-averse nature are also factors that need to be taken into account. It is, of course, possible to adopt a broad and anticipatory interpretation of what constitutes a credible threat, such as the one by David Luban, who believes that a threat is credible when its propensity for future armed attacks is clearly based upon characteristics such as 'militarism, an ideology favoring violence, a track-record of violence to back it up, and a build-up in capacity to pose

---

49 Following this philosophical tradition, the US Supreme Court has adopted this perspective by saying that: 'the people (...) erected their Constitutions, or forms of government (...) to protect their persons from violence' (Calder v. Bull, 3 U.S. 386, 388 (1798)) and that 'the obligation of the government to protect life, liberty and property against the conduct of the indifferent, the careless and the evil-minded may be regarded as laying at the very foundation of the social compact' (City of Chicago v. Sturges, 222, U.S. 313, 322 (1911)).

50 Allen Buchanan and Robert Keohane share the same opinion. They wrote: 'Adherence to the [current legalist approach] is too risky, given the widespread capacity and occasional willingness of states and nonstate actors to deploy weapons of mass destruction covertly and suddenly against civilian population' (2004, p. 3).

a genuine threat' (2004, pp. 230–231). When the international community faces such a threat, he believes we are justified in resorting to preventive measures against it[51]. In this case, it also means that such measures should have been taken against the Soviet Union from 1945 to 1949 or against North Korea before both countries were able to develop nuclear weapons and the means to use them against the rest of the world. In fact, alongside the aforementioned terrorist threat, the nuclear menace from a rogue state is also a very good example of the shortcomings of the pre-emptive logic. Indeed, thanks to intercontinental missiles or first-strike weapons such as submarines, rogue states (that show no concern whatsoever for human rights and display an aggressive rhetoric) now have the possibility to fire their missiles of death on their targets with little or no warning in a matter of less than half an hour. In fact, Manhattan Project scientist H.D. Smyth once described this weapon as being 'so ideally suited to sudden unannounced attack' (1945, p. 134), while Caryl Haskins called it 'an ideal weapon for aggressors' and Robert Oppenheimer 'a weapon of surprise' (Freedman & Michaels, 2019, pp. 55–56). As is now the case with contemporary terrorism, this daily reality of the Cold War made the distinction between prevention and pre-emption rather vague (Strachan, 2007, p. 36), while Walzer admits that those who developed the logic of the pre-emptive attack and opposed the idea of preventive strikes centuries ago 'did not take into account weapons of mass destruction or delivery systems that allow no time for arguments about how to respond' (2004, p. 147)[52]. This is why 'preventive-war thinking was surprisingly widespread in the early nuclear age', more specifically 'the period from mid-1945 through late 1954' (Trachtenberg, 2007, p. 43)[53]. Indeed, it has

---

51 For him, 'preventive war may be justified against a rogue state (in the sense given here, a threat state) aiming to construct WMD (in the sense given here, weapons that can cause mass casualties through a single use), if the state's intentions are hostile, because if the state succeeds in constructing WMD it may be too late to forestall a genocidal attack' (2007, p. 190).

52 Owing to that, Walzer wonders that 'Perhaps the gulf between pre-emption and prevention has now narrowed so that there is little strategic (and therefore little moral) difference between them' (2004, p. 147).

53 As Marc Trachtenberg explains, 'A whole series of major figures were very worried about what would happen if matters were allowed to drift and nothing done to prevent the USSR from building a nuclear force. They wanted the United States to do what it had to in order to prevent the Soviets from building such a force. They wanted America to bring matters to a head with the Soviet Union before it was too late. To me, it was just astonishing how many people were thinking along those lines – scientists, mathematicians, and philosophers (like Leo Strauss, John von Neumann, and Bertrand Russell), leading journalists and major political figures (including a number of senators), and, above all, a whole series of high-ranking military (and especially Air Force) officers. Even distinguished diplomats like George Kennan and

been reported that in 1954, there were talks at the US National Security Council level regarding the feasibility of a strike against the Soviet Union (Freedman, 2003, pp. 119–120). It was also not surprising to see a similar reasoning in the 1990s when it became clear that North Korea was actively trying to develop a nuclear arsenal, as key members of the Clinton administration considered launching a pre-emptive strike against the country's nuclear facilities before an agreement was reached. After all, is there a threat more imminent than a nuclear apocalypse that can fall on our heads within the next 30 minutes? Such a prospect has led the international community to adopt a more liberal understanding of pre-emption and imminence. This was, for instance, the case during the 1962 Cuban Missile Crisis when President Kennedy ordered the quarantine of the island even though a potential attack against the US was not imminent at the time. Despite the fact that the decision did not meet the conventional standards of a pre-emptive action, it was praised by the international community as being a 'cautious, limited, and carefully calibrated' response (quoted in Rockefeller, 2004, p. 134)[54].

We can, of course, argue about the risks associated with such an approach, but we also need to consider the chances that this rhetoric and weapon development may actually result in their use. In this regard, there is a fundamental difference between a state actor – even a rogue one – and terrorist organizations, namely the fact that the former is a risk-averse entity. Indeed, because of their territorial nature, states – even rogue ones – remain perfectly aware that a nuclear strike on their part would very likely result in self-destruction through nuclear retaliation. This is why Tony Coady wrote the following:

Charles Bohlen seemed to think that it would not have been too bad if war with the USSR had broken out before that country had developed a large nuclear force. And not just Americans: a number of leading European political figures were also thinking along these lines. Winston Churchill, for example, argued repeatedly in the late 1940s that matters needed to be brought to a head with the Soviets before it was too late, while the United States still enjoyed a nuclear monopoly. And Charles de Gaulle told an American journalist in 1954 that the "United States made a great mistake by not pursuing a policy of war" when it still had a "definite atomic lead"' (Trachtenberg, 2007, p. 43).

54 In his 22nd October 1962 address to the nation, he said the following: 'Neither the United States of America not the world community of nations can tolerate deliberate deception and offensive threats on the part of any nation, large or small. We no longer live in a world where only the actual firing of weapons represents a sufficient challenge to a nation's security to constitute maximum peril. Nuclear weapons are so destructive and ballistic missiles are so swift that any substantially increased possibility of their use or any sudden change in their deployment may well be regarded as a definite threat to peace'.

> Another nation's development of weapons, including weapons of mass destruction, may create various worries and uncertainties, but there is so much that can come between that development and its hostile use, that we should not risk the hazards of war on behalf of the alarming prediction. (...) Moreover, the existence of rapid delivery systems does not eliminate the space for arguments about response since these arguments will have been canvassed, and policies put in place, long before an attack. Widespread knowledge of military potential has become as swift and comprehensive as the speed of delivery systems though fallibility still attaches to both. In addition, those with, or developing, such weapons are increasingly alert to the risks they face from preventive military action (especially when it is loudly advocated) and take steps to make it more difficult for enemies to destroy their facilities with speed and precision; hence awareness of the likelihood of preventive military attacks increases the probability that they will be thwarted (Coady, 2013, p. 194).

However, this is not the case with terrorist organizations; because of their non-territorialised nature and fanaticism, they do not have the same sensitivity to the consequences of their actions. In this sense, if there are reasons not to exaggerate such threats from state actors, there are also reasons not to show the same optimism with terrorist groups. This is not to deny that terrorist groups have in the past showed an aversion to unnecessary risk. However, what needs to be emphasised is that their motivation in this regard has not been the fear of retaliation but rather their fear of not succeeding. This is why their threats should not be minimised in the same way as those made by state entities[55].

There are, therefore, reasons to believe that the way the pre-emptive logic is conceptualised poses problems in the context of contemporary terrorist organizations and that it ought to be amended to enable states that are threatened by these groups to use more proactive measures to defend themselves and protect their citizens (Schmitt, 2003; Franck, 2002). Thus, we are obligated to abandon the notion of imminence[56] and to go beyond the temporal proximity of a threat by thinking about a new threshold that will need to be crossed before an anticipatory act of self-defence can be justified since terrorist attacks will

---

[55] From this perspective, it is difficult not to agree with the 2002 US National Security Strategy, which argued that 'today's enemies are no longer status-quo oriented, risk-averse states, but "rogue" states that are prepared to use their capabilities actively rather than simply defensively, or non-state actors who are similarly averse to recognizing limits on their behavior, and who are prepared to use extreme tactics, including the mass killing of civilians, and to sacrifice themselves in doing so' (quoted in Brown, 2013, p. 34).

[56] Owing to the way an imminent threat is defined, sticking to this notion is a dead end as it can only be understood in a restricted way which, as a result, makes states impotent against the nature of today's terrorism.

never meet the former standard[57]. This void can be filled with the already mentioned notion of the 'credibility of a threat', which can be highlighted with the two previous examples of offenders threatening customers in a bar. This course of action ought to be used solely against groups whose threatening rhetoric matches their actual capacity to transform their promises into reality. This means that an organization that promises to drop a nuclear weapon in a crowded urban area but that does not have the capacity to do so should be considered similarly to the drunk customer in a pub and, accordingly, should not be harmed in the same way as a group with the genuine capacity to fulfil this threat. Of course, a history of indiscriminate violence by these groups would simply reinforce the credibility of their threat[58]. In hindsight, this criterion is by no means a conceptual novelty as it is already a constitutive element of the R2P principle and has been used to justify humanitarian interventions in the past. This is what the next section will show.

3.2 R2P and the Credibility of a Threat

In order to effectively protect states and people from some contemporary organizations, I argue that the notion of imminence has to be replaced with one that revolves around the credibility of the threat. While this can be troubling for some, who might consider this criterion too generous, it has to be mentioned that it is a cornerstone of the R2P principle and previous humanitarian interventions. In fact, this idea was used quite successfully in 1991 against Iraq following the liberation of Kuwait. Soon after the Desert Storm Operation, a civil war ignited in the northern and southern parts of Iraq after the Shiite Muslims and the Kurds launched military operations against the Iraqi forces, which remained loyal to Saddam Hussein. This was mainly the result of President Bush's explicit encouragement to the Iraqi people to overthrow their dictator,[59] which led these

---

57 In this regard, Daniel Brunstetter and Megan Braun's approach to *jus ad vim* is unsatisfactory as they justify its use only when threats from terrorist groups are imminent (2013, p. 96–97). They do not consider that this notion hardly applies to the *modus operandi* of terrorist organizations.
58 Although I see this criterion as being important, it is not required in order to justify measures short of war against terrorist groups whose means of actions match their rhetoric. Otherwise, this would require us to first sacrifice the lives of innocent civilians before striking against them.
59 In his remarks in front of the American Academy for the Advancement of Science on 15th February 1991, President Bush said the following: 'But there is another way for the bloodshed to stop, and that is for the Iraqi military and the Iraqi people to take matters into their own hands to force Saddam Hussein the dictator to step aside and to comply with the UN and then rejoin the family of peace-loving nations' (quoted in Malanczuk, 1991, p. 117).

two groups to believe that they would be supported by the US forces if they chose to fight. However, as it turned out, President Bush was only offering moral support[60] and soon after attacking Iraqi forces and taking control over garrisons in early March 1991, the momentum quickly shifted. This was made possible despite the fact that Saddam Hussein's military forces had suffered a crushing defeat, because the Iraqi dictator had been able to prevent his elite Republican Guard units and a large number of his tanks from being captured or destroyed by the coalition forces by agreeing to a ceasefire about 100 hours after the beginning of the US-led offensive to liberate Kuwait. Equipped with helicopters, combat aircraft, and heavy tanks, the Iraqi forces retook control of their lost territories. In the north, the counter-offensive led to a massive refugee crisis when about one million Kurds fled into the mountains, fearing for their lives. Their reaction was triggered by the Anfal genocide, to which their people had been victim for nearly 30 years. In its earliest stage, the policy, which was also known as 'Arabisation', was mainly oriented at displacing Kurdish families from the region to other areas of Iraq where they were monitored by the Iraqi military and given minimal food and water, which led many to their deaths. In return, with the clear aim of altering the demography of the northern part of the country, poor Arabs were encouraged to move to the abandoned houses of these families. This policy was also associated with mass murders of men of military age and the detention in harsh conditions of Kurdish women and children. Over this period, nearly one million Kurds were estimated to have died. However, the most remembered event of this genocide certainly remains the Halabja chemical attacks in March 1988, when the Iraqi regime resorted to the indiscriminate mass gassing of an entire town through the use of sarin, VX, and mustard gas that killed at least 5,000 and injured around 7,000 civilians. With little food as well as inadequate shelter and clothing for the harsh conditions of the cold mountains, it was estimated that between 1,000 and 1,500 refugees were dying on a daily basis. Faced with the distress of the Kurds, the West decided

---

60 The insurgents were most probably victims of the *raison d'état*. As Peter Malanczuk wrote: '(...) in view of American interests in the region as a whole, it did not mean support for a division of Iraq in the wake of the Shiite insurrection in the south and a corresponding Kurdish uprising in the north. On the contrary, the territorial integrity of defeated Iraq needed to be secured in order to preserve Iraq's function as a balance, primarily against Iran. Although the Islamic Republic did not say so officially, it was clear that only Iran had an interest in a successful Shiite revolution in the south of Iraq. The establishment of an independent Kurdistan in the north of Iraq, on the other hand, not only would have raised the issue of control over the important oil resources in the area, but also would have posed a threat to the security of neighbouring States, in particular Turkey' (1991, pp. 117–118).

to act. On 7th April, the US began Operation Provide Comfort by airlifting food and medicines to the refugees, and the following day John Major, the British Prime Minister, asked for the creation of a safe zone in the area that would provide security to the Kurdish people, an initiative President Bush finally agreed to a few days later. This latter action, which led to the creation of a no-fly zone above the 36th parallel that remained effective until the 2003 invasion, effectively cut off Saddam Hussein's regime from the northern part of the country and allowed the Kurds to be protected from further exactions. In the 12 years that followed its creation (as well as the no-fly zone in southern Iraq created in August 1992 to prevent human rights violations against the Shiite Muslims), many limited interventions occurred either when the Iraqi air forces attempted to breach the exclusion zone or when the Iraqi armed forces showed aggressive intent. For instance, in September 1996, 44 cruise missiles were launched against Iraqi air defence targets as retaliation after Saddam Hussein had launched an offensive in Iraqi Kurdistan. In September 1992, an American F-16 shot down an Iraqi Mig-25 that was caught flying inside the southern no-fly zone. Alongside these restrictions, following the adoption of Resolution 687 by the UN Security Council, Iraq also had to submit itself to a series of sanctions that prevented Saddam Hussein from using, developing, constructing, or acquiring chemical, bacteriological, or nuclear weapons. Iraq was also limited in its capacity to import and export goods. The objectives of these sanctions were to prevent the mass murder of Iraqi civilians as well as another invasion of Kuwait and to eliminate Iraq's capacity to use WMD.

As argued by former Marine Corp Commander General James Jones, this intervention proved to be 'one of the greatest American military operations of the 20th century' (2017). Through its soft approach, the US-led coalition was not only able to provide effective humanitarian aid to a vulnerable population but also allowed the Kurds to return home and live peacefully until the 2003 invasion. Moreover, the effectiveness of the second set of sanctions cannot be denied. Despite the claims made by the Bush administration prior to the 2003 invasion, Saddam Hussein's regime did not produce WMD following its defeat against the coalition forces in 1991 after the liberation of Kuwait. This is why Walzer argues that the effectiveness of this containment system made the war of 2003 unnecessary (2006, pp. xiii–xiv).

While this is undeniably true, we must admit that the imposition of no-fly zones was illegal under international law. Indeed, preventing the Iraqi air forces from operating on a significant part of its national territory and being the victim of surgical strikes were clear breaches of Article 2, Paragraph 4 of the UN Charter. This was clearly stated in April 1991 by the UN Secretary-General, who perceived the intervention as a 'serious, unjustifiable and unfounded attack on the sovereignty and territorial integrity of Iraq' (Malanczuk, 1991, p. 124). On the contrary, and similar to what Walzer has said, a lot of people would argue that there is little room to pretend that this intervention was not justified from a moral perspective. I believe that most people would come to this conclusion because of Saddam Hussein's proven lack of consideration for human rights – with the Kurdish genocide being the most explicit example. From this perspective, the resort to the aforementioned forceful measures that limited Iraq from enjoying full political sovereignty could be seen as being the precursor to what became known in the 21st century as the R2P principle. This means that the success of such humanitarian interventions has more to do with the logic of preventive than pre-emptive wars since the notion of imminence cannot always be assessed in the same way as when we are facing a conventional conflict between two states, as in 1967 with Israel and its Arab neighbours. To prevent civilians from being massacred, it might be required to assess imminence in a manner that places a higher value on conjectures and intentions than on clear evidence of an upcoming attack.

The case of Iraq in 1991 is a good example in this regard. Intuitively, it is easy to justify the restrictive actions imposed on Iraq based on the need to prevent civilians from a genocide. In fact, this rhetoric was used by political leaders such as Masoud Barzani, leader of the Kurdish Democratic Party, who called on the international community to help stop the genocide against the Kurds, or Germany's Foreign Minister, Hans-Dietrich Genscher, who described on 5th April 1991 the actions of the Iraqi military as akin to a genocide (Malanczuk, 1991, p. 119). This opinion is shared by analysts such as Samantha Power (2002, p. 241), who have seen in Operation Provide Comfort an unprecedented intervention that marked the beginning of a new post-Cold War era in terms of genocide prevention. But, at that time, there was no empirical evidence of an ongoing violation of human rights against the Kurds or of any planning to pursue actions similar to the ones in Halabja a few years before. Of course, and as said before, many of them were dying on a daily basis in the mountains of northern Iraq, but this was the result of the civil war itself and not the direct exactions of Saddam Hussein's men. If the interventions by the US-led coalition were based on a humanitarian perspective and for genocide prevention, they were not the result of a clear and imminent threat that the Iraqi government had resumed or was about to resume its mass murder policy against the Kurds despite the rhetoric used at that time by those who were advocating for an intervention. Accordingly, what is striking about the 1991 sanctions is their preventive nature as a possible counter-attack against Kuwait or the coalition forces with the use of chemical or nuclear weapons was not imminent at all. The same can be said regarding genocidal actions against the Kurds. The limitations on Iraq's sovereignty were mostly the result of a fear that Saddam Hussein may have resumed his previous deeds in the aftermath of his defeat. At no point was the international community certain that this would happen or that the killing of members of this minority group was imminent. The 2011 crisis in Libya also showed how unpredictable human rights violations can be. As said by Ruben Reike, 'It is interesting to note that at the time that the first protests erupted, Libya was not considered to be at risk of mass atrocities or violent conflict by any of the various risk assessment mechanisms' and 'Despite Gaddafi's notorious record on human rights, the country was widely seen as relatively stable' (2012, p. 126).

In light of all this, why is it that the decisions to intervene in Iraq and Libya seemed appropriate from a moral perspective? I would argue the statements made by the Iraqi and Libyan regimes as well as their capacity to fulfil their threats made these fears credible and justified acting against them. This point is, I believe, the cornerstone element of a revised view of what constitutes an imminent threat in light of the new nature of contemporary political violence. We cannot ignore the fact that, in a speech made on 16th March, Saddam Hussein said that Iraqi armed forces would crush the Kurds with all possible means, including the use of chemical weapons. The fact that Iraq had previously shown its utter lack of respect for the Kurds and its willingness to use this sort of weaponry against them simply made that threat even more credible. Moreover, despite its quick defeat against the coalition forces, the Iraqi military still had at its disposal the necessary equipment to resume its genocidal campaign against the Kurds, outweighing them with heavy tanks, helicopters, and planes. If we were to follow the logic of the pre-emptive attack, we would have had to wait for Iraq to manifest a degree of active preparation against the Kurdish population before acting against the regime. However, because these attacks can be planned covertly and performed very quickly, standing by the conventional logic of pre-emptive war would have meant waiting for the Iraqi military to resume the gassing of its Kurdish minority before taking action against Saddam Hussein. If the price to pay is waiting for the start of the actual murder campaign before acting, the R2P principle simply loses its purpose and would bring us back to situations such as those in Rwanda or Bosnia and Herzegovina in the 1990s.

In a context where imminence cannot be assessed and when civilian lives are at stake, the intentions of an actor with the means to fulfil its ambitions are sufficient criteria for the justification of actions aimed at preventing the menace from ever being carried out[61]. In theory, applying these criteria would have allowed the international community to prevent the Rwanda genocide. If there

---

61 As Emer de Vattel wrote in the 18th century, 'power alone does not threaten an injury: it must be accompanied by the will' (Book 3, Chapter 44).

were no signs that allowed the peacekeepers on the ground to foresee the genocide hours or days before it actually started on 7th April 1994, their commanding
officers, nonetheless, had evidence that it was going to happen at some point.
Indeed, as reported by General *Roméo* Dallaire – who was leading at the time
the UN Assistance Mission for Rwanda – in his book *I Shook the Devil's Hand*
and in many of his interviews, it was obvious many months before the beginning
of this humanitarian tragedy that a genocide was being prepared by some Hutus.
On many occasions, he sounded the alarm. On 20th January 1994, the Canadian
general was warned by an informant that radical members of the Hutu government were planning to eliminate the Tutsi. More specifically, he was informed
about the existence of four separate weapons caches in Kigali filled with ma-
chetes, AK47s, and grenades, revelations that were confirmed during the second
week of February when another informant revealed similar information[62]. Even
though the prospect of a genocide in Rwanda never met the conventional criteria
of imminence that would have legally justified pre-emptive measures, it was
clear months before it was triggered that the radical Hutu had the means to fulfil
their intentions of eliminating whom they called the 'cockroaches'. Yet, no one
listened to General Dallaire, which led to one of the most dramatic humanitarian
tragedies of the 20th century (Caron, 2018b, 2019b).

This logic bears a lot of similarities with the way Randall R. Dipert has defended the resort to some preventive wars. For him, when a threat is substantial
and unjustified, and waiting for the potential attacker to strike first poses a serious risk, there are grounds to justify an anticipatory attack. He gives the example
of a belligerent neighbour who has previously attacked you and your family
without reason and has sworn to do it again. Thanks to the mailman who has
informed you that the man has been delivered hand grenades, a machine gun,
and an RPG, you know that his threats can be fulfilled. Although you do not
know precisely the day and time of his attack, Dipert argues that these facts
would certainly constitute legitimate grounds for a preventive attack (2006,

---

62 General Dallaire sent cables to Kofi Annan, then head of the UN Department of Peacekeeping
Operations, about this information and that his informant had been ordered to register all Tutsi
in Kigali, which he suspected was a way to facilitate the future killings (…). He was also horrified
to hear the propaganda of Hutu extremists echoed in the media, notably the newspaper *Kangura*
and the *Radio Télévision Libre des Milles Collines*, calling explicitly for the elimination of the
Tutsi (Thompson, 2007, pp. 1–12). This is why General Dallaire repeatedly asked his superiors
for more freedom with regard to the UN mandate that did not allow him to disarm the militias.
He also sought to block the Hutu radio transmissions and asked for the sending of more UN
peacekeepers. However, all his demands were refused and his worst fears became a reality in
April 1994.

pp. 37–38). It is certainly hard to disagree with him. However, through this example, Dipert is unconsciously adding another criterion to his list, that is, the fact that your neighbour possesses the means to fulfil his threats against you. This element is, in my view, required if we are ever to justify anticipatory attacks that currently fall outside the current logic of pre-emptive war. Relying solely on Dipert's first three criteria is not sufficient as they would allow states to attack any threatening entity – irrespective of its capacity to actually do so. In hindsight, this view could lead to a generalisation of political violence and needs, therefore, to be complemented with this last point.

Evaluating the imminence of an attack by a terrorist organization is, of course, a complex matter and following the conventional logic of the pre-emptive attack is simply not an option. More precisely, terrorist organizations generally strike when their enemies least suspect it, so as to instil terror within the civilian population. Contrary to states resorting to a conventional attack against their enemies, namely through the use of tanks, planes, and infantrymen, these terrorist organizations can remain hidden until the moment they choose to act. Thus, the imminence of such attacks can never be assessed by the targeted state. As said before, there is a huge difference between a state showing an aggressive rhetoric by reinforcing its military with conventional weapons and a terrorist organization – or a state sponsoring terrorism – that is also trying to develop WMD displaying the same intentions. If in the former case, the development of the threat can be evaluated through the deployment and mass mobilisation of troops, it cannot be done in the latter case. Unfortunately, the current state of international law leaves no lawful options to states targeted by the latter threat to defend themselves. This is obviously problematic as it only leaves two options: inaction – which can lead to the deaths of thousands of civilians – or resorting to illegal actions.

For the sake of saving civilians from being massacred, actions that fall outside the scope of the category of the pre-emptive attack and that have more in common with preventive attacks may be necessary. It has to be noted that this logic is being used in many societies against individuals who do not pose an imminent threat according to the conventional understanding of pre-emptive attacks. This is the case with pre-trial detention, which prevents individuals accused of a crime from being released on bail because they are deemed likely to commit further crimes if released. In other cases, like in Canada, some individuals who are labelled 'dangerous offenders' may be subjected to indeterminate prison sentences irrespective of whether they have already completed their original sentences. In such cases, the court has to take into account various factors, such as if the convict has demonstrated a repetitive pattern of violence that is likely to persist upon release[63], a total indifference to the consequences of his or her actions, and the brutal nature of the crime for which he or she has been convicted, which shows an unwillingness to abide by normal standards of behaviour. In these cases, the imminence of the threat is not the same as in a situation that justifies a pre-emptive attack. It cannot be evaluated in light of the individual's active preparation to commit a crime in the near future, but rather to the likelihood that he or she may once again break the law and harm others.

This was the logic behind the decision to impose sanctions on Iraq after 1991. Despite his defeat, Saddam Hussein did not show contrition and a willingness to transform his country into a law-abiding political entity. On the contrary, he remained a credible threat to Kuwait. For instance, in 1994, he ordered the transport of ammunition and support material as well as two divisions of his elite troops to move as close as 12 miles from the border with Kuwait before withdrawing them when the US responded by mobilising more troops in the region. In other words, because of its unwillingness to change, Iraq became, as Walzer puts it (2002), 'a state on parole' that could not be trusted by the international community and was seen, rightfully or not, as a serious menace to others similar to an individual labelled a dangerous offender.

These types of preventive measures are a constitutive and recognised element of states' most basic obligation – to protect their citizens – which also ought to apply in the case of terrorism. In light of what has already been said, it is needless to recall that an international terrorist organization that refuses to apply the principle of discrimination poses a direct threat to this duty as it '[threatens] the dignity and security of human beings everywhere, endangers or takes innocent lives, creates an environment that destroys the freedom from fear of the people, jeopardizes fundamental freedoms, and aims at the destruction of human rights' (Office of the United Nations High Commissioner for Human Rights, p. 7). This is why it is legitimate for states not only to have an effective criminal justice system that deters the commission of offences and punishes those who have attempted to infringe on other people's natural rights but also to take preventive measures that will protect individuals whose lives are known or suspected to be at risk: a situation that includes a terrorist threat. This positive obligation has been reiterated in numerous court judgments (see Kilic v. Turkey, 2000; Velasquez Rodriguez v. Honduras, 1988; Delgado Paez v. Co-

---

63 For instance, an individual found guilty of a sexual offence for a third time is deemed unlikely to control his or her sexual impulses, and the likelihood of that behaviour being repeated is very high.

lombia, 1990). Of course, in light of the complexity of modern societies, the whole challenge is to determine when this duty arises. The answer to this enigma bears a lot of conceptual similarities with what has been described earlier as a credible threat. Indeed, for the European Court of Human Rights:

> For a positive right to arise, it must be established that the authorities knew or ought to have known at the time of the existence of a real and immediate risk to the life of an identified individual or individuals from the criminal acts of a third party and that they failed to take measures within the scope of their powers which, judged reasonably, might have been expected to avoid that risk (2000, par. 63).

This judgment resulted from the murder of a journalist named Kemal Kilic who had been working for a pro-Kurdish newspaper and who had informed the state authorities about death threats made against the employees responsible for its sale and distribution. Moreover, he had informed them about known cases of employees who had been attacked and killed. The authorities chose to ignore these claims and Kilic ended up being shot dead a couple of weeks later. The European Court found the Turkish authorities liable for not having lived up to their obligation to protect Kilic and the other employees of the newspaper. In this case, as it was a well-planned surprise attack that cost Kilic his life, the notion of imminence did not play a role in the court's decision. What mattered was rather the credibility of the threat ('the authorities knew or ought to have known at the time of the existence of a real risk to the life of an identified individual or individuals from the criminal acts of a third party'). As the consequences are similar, the same liability may be applied against states that know about a terrorist group that has the means to fulfil its promises of destruction and murder against innocent civilians. In this case as well as when it comes to the R2P principle, it is the credibility of a threat that triggers a state's positive duty to protect its citizens through proactive and preventive actions. If this is a valuable option when it comes to humanitarian interventions, it should be the same against contemporary terrorist organizations that pose a similar risk to innocent civilians. Accordingly, because of the contemporary challenges the international community has to face, namely the obligation to protect civilians and the nature of terrorism, there is a need to rethink the legal framework set in place after WWII by widening the spectrum of acceptable political violence that falls short of war. However, with this in mind, we now need to determine what ought to be these means of actions that will prevent such a threat from becoming a reality. This is what the next chapter will analyse.

Chapter 4:
Thinking About Violent Alternatives to War

Following what has been said previously, the threat of contemporary terrorism requires us to rethink the threshold of legitimate violence under new criteria. As terrorist attacks will rarely appear as imminent as those of state actors and because the former are not as risk-averse as the latter, it has been suggested that violent measures ought to be taken insofar as the threats posed by these groups are deemed credible. That said, it becomes necessary to understand precisely the types of actions that can be taken against terrorist organizations and entities harbouring or protecting them as well as the moral criteria that ought to surround their use. This is what this chapter will explore.

4.1 The Differences Between Fighting Terrorist Organizations and States Sponsoring Terrorism

First, we need to admit that NVATW may still be considered valuable options in cases of terrorist groups benefitting from the support of state actors. In such cases, we cannot deny that economic or diplomatic sanctions or other forms of 'soft war' may play a fundamental role in forcing a state to cut its relations with these groups. Achieving this goal ought to be the primary objective of states that are being threatened by the terrorist organization, as it can contribute to eliminating the threat without having to resort to violent solutions. Indeed, when terrorist organizations have to operate on their own, it might be more difficult for them to acquire WMD or have access to a safe haven where they can train and plan their future attacks. When this is the case, their threats may simply be a matter of rhetoric. On the contrary, the close collaboration of these two entities is highly problematic and should not be ignored. Indeed, this collaboration may also be an effective way for rogue states to wage indirect war when they know that a direct confrontation with their enemy would most likely not go in their favour. Moreover, by not being the direct perpetrators of a terror attack, rogue regimes may either plead ignorance or claim a lack of responsibility and actually manage to get away with it by avoiding any direct retaliation against them, while a state's protection may also prevent terrorists from ever being prosecuted.

In such a case, Walzer's assessment of the containment system imposed against Iraq in 1991 not only provides insight into the types of actions that may be used against these states, their effectiveness but also sheds light on

https://doi.org/10.1515/9783110729894-006

how our understanding of pre-emption is actually a reality of world politics. For Walzer, if Saddam Hussein was a threat to regional stability and the Sunni and Kurd populations, overthrowing him through a full-scale war had detrimental consequences on both aspects. Indeed, the 2003 invasion not only caused a major humanitarian crisis that ultimately resulted in the deaths of tens of thousands of civilians but also led to sectarian clashes and inter-communal violence and the development of a new terrorist organization (ISIS) that sparked violence and instability outside Iraqi borders. He cannot help himself from thinking that the severe containment measures implemented against Saddam Hussein in 1991 proved way more effective at preventing the Iraqi dictator from killing civilians and producing WMD than waging a full-scale war against him and his regime. More precisely, the military measures imposed on Iraq – namely the no-fly zones in the northern and southern parts of the country – most probably saved the lives of people who (as we can assume since the threat was credible) would otherwise have been mercilessly butchered by the Baathist regime. Furthermore, the embargo that prevented Iraq from importing weapons and the inspection system put in place by the UN to prevent Saddam Hussein from acquiring WMD turned out to be more effective in keeping the regime under control and preventing it from violating international law. Walzer refers to these measures as being constitutive of the *jus ad vim* category and as 'measures short of war'. Not only can they be highly effective in preventing war, human rights violations, and terrorism but they were also able to prevent the unpredictable consequences of war that can hardly be contained even when states are fully committed to respecting the moral rules of warfare.

As already argued in the introduction of this book, there are reasons to question the necessity of adding a fourth category to the Just War Theory alongside *jus ad bellum, jus in bello*, and *jus post-bellum* since measures short of war are already essential components of *jus ad bellum*. Indeed, if war is thought to be the last option, creating this fourth category would be, to borrow Helen Frowe's expression, simply redundant (2016). However, I argue that this is a false debate caused by Walzer's inability to distinguish between NVATW and VATW, as NVATW are undoubtedly essential components of the *jus ad bellum* category. In this case, many of the containment measures used against Iraq in 1991 fell within that category, namely an economic embargo and the monitoring by international observers of the country's compliance with its obligation not to acquire or develop WMD.

The resort to these NVATW can also be used against states suspected of sponsoring and harbouring terrorist organizations. These measures can prove efficient at preventing these organizations from ever being able to fulfil their promises of destruction if they are deprived of their capacity to acquire WMD thanks to international surveillance. The idea behind such sanctions is making the targeted state realise that pursuing this course of action is a political dead end that will most likely be detrimental to the survival of its regime. If this is achieved, the risks of these organizations' threats ever becoming credible are slim to none, which eliminates the necessity of setting down the path of VATW. However, this alternative may be entertained under certain conditions. For instance, as highlighted by Walzer, the success of NVATW implies the willingness of the international community to enforce them. When this is not the case and when a power that has the capacity to provide the necessary assistance in the development of WMD or to deliver them to a rogue state starts collaborating with terrorist organizations, the effectiveness of NVATW is questioned as the effects of these sanctions can be bypassed thanks to this ally. This resembles a situation where a child grounded by his mother and sent to his room without dinner is secretly brought food by his father. The child, therefore, has no incentive to change his behaviour and his mother's efforts run the risk of being pointless.

The resort to VATW may also be considered when NVATW are not producing their expected effects, leading the statesmen of rogue entities to be insensitive to international reprobation. This means that there are reasons to believe that the incapacity to change their country's course of action will ultimately result in a situation where terrorist organizations will have the possibility to pose a credible threat to innocent people. This will also be the case 'when it is already too late', namely a situation where a rogue state is already in possession of or about to acquire WMD. In this case, the capacity of a terrorist organization to get its hands on such a weapon and to transform its threat into reality becomes real. The threat is no longer a matter of rhetoric but has rather become credible. When such situations occur, violent pre-emptive actions may be considered both against members of the terrorist organization and the rogue state, such as – for the former – the assassination of those posing a threat or – for the latter – destruction of the facilities used in the fabrication of WMD. In such cases, waiting for the threat to become imminent is not an option for the reasons already discussed.

Obviously, the first option ought to be considered because there is no other – or very few or dangerous – non-violent alternatives at the disposal of states under threat because of the close collaboration between terrorists and the authorities of a rogue state. Indeed, in the case of terrorists operating on a state's territory without benefitting from the support of its authorities, as was the case in Mali with Al-Qaeda in the Islamic Maghreb, non-violent alternatives can be entertained. For instance, when the international community can benefit from the state's support, it greatly facilitates the development of a collaborative action plan with the local military and police forces, which makes options such as arresting and bringing these individuals to trial conceivable. However, when these individuals benefit from the protection of the state from which they are operating, such collaboration is futile as it might simply make these terrorists realise that they are under surveillance and accelerate their plans to disappear. This is why striking against them violently and by surprise may be the most proportionate solution.

In the same vein, VATW may also be entertained when a state that is not in any way collaborating with a terrorist organization posing a credible threat shows its unwillingness to collaborate with the international community. In fact, considering the devastating potential of terrorism with regard to civilians' lives, I would argue that preventing their mass murder bears no inherent differences from states' obligation to prevent their citizens from being the victims of genocide, war crimes, ethnic cleansing, or crimes against humanity, also known as the R2P principle. According to this rule, states have the primary responsibility to prevent these crimes from occurring on their territory and the international community has the obligation to encourage and assist states in this regard. However, if a state is manifestly failing to protect its population because of its inaction or because it is behind these crimes, the international community must be ready to take the appropriate steps, including resorting to military actions. If we are to believe that the murder of innocent civilians through the use of a nuclear, bacteriological, or chemical weapon used in downtown New York, Paris, London, or Prague is similar to the systematic killing of people in crimes forbidden under the R2P rule, then states also have a 'responsibility to prevent terrorism' (R2PT). Thus, it is the obligation of states that are unable to respect the R2PT principle to seek external help in order to prevent terrorist groups from enjoying the benefits of having a base of operations. Failure to do so or an unwillingness to seek this sort of help – because the state authorities are actually supporting these organizations – would require action on the part of the international community (Bethlehem, 2012).

It must be noted that this normative view is very close to the current international norm for interpreting state responsibility for actions committed by non-state actors (Graham, 2010). This has been one of the most significant shifts since 9/11. Prior to these events, a state's liability for the actions of a third party was based on the 'effective control' criterion. This resulted from the International Court of Justice decision regarding the US' responsibility for the actions of the *contras* in Nicaragua. The judges concluded that in order to be held liable for the actions of non-state actors, it had 'to be proved that that State had effective control of the military or paramilitary operations in the course of which the alleged violations were committed' (Nicaragua v. United States, 1986, par. 115). This idea was later confirmed by the Appeals Chamber of the International Criminal Tribunal for the former Yugoslavia, which said that '(...) when private individuals carry out acts contrary to international law, the only way to attribute such acts to the host-state is to demonstrate that the State exercises control over the individuals' (Proulx, 2005, p. 106). However, following the 9/11 attacks and because of the relationship between the Taliban government and Al-Qaeda[64], states' liability for acts of terrorism has been broadened in such a way that the new international norm now centres around the idea that states can be held responsible for having failed to prevent an attack by a terrorist group operating from their territory. Thus, states are not only responsible for their direct breach of a treaty or of customary law but also for their failure to prevent these violations from occurring. Accordingly, 'this clearly confirms the right of a victim state to treat terrorism as an armed attack and those that facilitate or harbor terrorists as armed attackers against whom military force may be used in self-defense' (Franck, 2002, p. 54).

All of this means that in a country where global terrorist organizations are operating without the collaboration of the former, its authorities have the obligation to fight this threat and, if they are unable to pursue that objective on their own, they must seek assistance from the international community in order to meet their R2PT. Concretely, this means that it would first be appropriate to think of assisting the local authorities in policing by providing them with weapons and military equipment, training, or even the services of a handful of highly trained elite soldiers for arrest scenarios. Whatever the form of this aid, it should be with the collaboration and approbation of the state's authorities. Having the capacity to benefit from this help may also provide targeted states access to better intelligence about the nature of the threat and its credibility. Thus, it may allow these states to minimise the resort to violence if this allows them to monitor these groups more closely. As an example, benefitting from the state's collaboration may allow a threatened state to consider working with local fiscal or bank authorities and to freeze the financial assets of this menacing group, thereby lowering its capacity to fulfil its deadly ambitions without

---

64 As mentioned by Vincent-Joël Proulx, 'Publicly available facts tend to demonstrate that the Taliban harbored terrorists and, at best, provided them with limited logistical support. However, it is difficult to contend that the Taliban government did in fact exercise effective control over Al-Qaeda; Al-Qaeda had a complex structure and much organizational and operational autonomy from the Taliban. The Taliban probably did not know of the 9/11 attacks beforehand and never endorsed them. Further, it does not appear that Al-Qaeda was acting as a *de facto* agent of the Taliban' (2005, p. 121). This meant that, according to the Nicaragua ruling, it would not have been allowed under international law to take measures against the Taliban authorities. This is why the military intervention in Afghanistan created a new precedent in international law.

having to violate another state's cyberspace. On the contrary, not benefitting from the collaboration of states where these groups are operating changes the range of options. In such cases, not having the support of the local police or intelligence authorities makes it extremely difficult for threatened states to keep a close eye on the terrorists' actions. Moreover, with the support of the state, members of these groups may be tipped off by informants working for the authorities who may also help them disappear by providing fake identities or new passports. When such a scenario arises, it is difficult to foresee any other alternatives than the lethal targeting of members of these terrorist organizations.

For its part, the targeting of the infrastructures of a state collaborating with terrorist organizations that are suspected to produce or be instrumental in the production of WMD can take many forms. The most famous case in this regard is certainly the highly controversial bombing of the Osirak nuclear reactor by the Israeli Air Force on 7th June 1981. Then Israeli Prime Minister, Menachem Begin, claimed that this was a justified case of self-defence. Indeed, there were serious doubts about the claim that Iraq's nuclear programme was being developed only for civilian purposes. This was a fear shared by the US Senate Committee on Foreign Relations, which went on record saying that there were 'serious misgivings regarding the ultimate character and direction of the Iraqi nuclear program' (Boudreau, 1993, p. 29). These fears were coupled with the fact that Iraq and Israel had been fighting various wars for almost 20 years and that Iraq was the most vocal and radical Arab state that was refusing to recognise Israel's right to exist. Furthermore, Iraq was also well-known at the time for its chilling rhetoric against Israel, with the most explicit being Saddam Hussein saying in the months that preceded the bombing that the reactor was intended to be used against 'the Zionist enemy' (Boudreau, 1993, p. 25).

Walzer's discussion of this case is interesting even though it falls short of providing a thorough and convincing explanation about why it was a justifiable action considering the circumstances under which it took place. As he writes, even though 'the Iraqi threat was not imminent' at the time, he believes it was nonetheless a 'justified preventive attack' because it involved the development or the delivery of WMD, which made it also, in a sense, a pre-emptive attack (2004, p. 147). As said previously, it is true that having WMD such as nuclear weapons tends to challenge our understanding of imminence since these weapons can be used to destroy one's enemies with little or no warning. This possibility is further reinforced by the rhetoric of states having them in their arsenal. In this sense, Walzer's assessment appears to make sense. However, even though the notion of imminence seems to have vanished when the use of such weapons is being openly entertained by states, it would be a mistake to ignore the broader picture and conclude that attacking one's enemies in similar circumstances is acceptable and a matter of self-defence. The impossibility of determining with certainty the imminence of a possible attack needs to be re- placed by another objective factor. I humbly argue that my criterion of a threat's credibility ought to be the much-needed *ersatz*. The question is whether belli- cose states that have or aim to acquire nuclear weapons really plan to use them against their enemies. Further, as argued in the previous chapter, the fact that state actors are territorial entities that cannot escape immediate retal- iation in the case of an attack on their part makes their rhetorical threats of fire and destruction sound not very credible. As written by Donald G. Boudreau in his discussion of the bombing of the Osirak reactor:

> It requires a tremendous leap in judgment to presuppose that a country's nuclear capability (and potential) translates into genuine intentions of launching a nuclear attack against an- other country. It is one thing for an Arab dictator [Hussein] to launch callously, in rank vi- olation of international law, chemical weapons against Iran and his country's own defense- less Kurdish population. It is quite another thing, however, for that same Arab dictator to plan earnestly launching a nuclear attack against a country possessing Israel's nuclear ca- pability – a country firmly allied under the United States' protective umbrella (Boudreau, 1993, p. 30).

By ignoring the notion of credibility, Walzer's assessment of the Osirak bombing, which he sees at the same time as a preventive and a pre-emptive attack, is prob- lematic. Striking first against a non-credible threat under the pretext that its im- minence cannot be assessed is clearly a preventive action that cannot be con- fused in any way with the revised view of pre-emptive attack that has been defended so far in this book. However, for reasons that were previously evoked, the risk is different when a similar state is actively collaborating with terrorist organizations; thus, things would have been different had there been evidence that Iraq was closely collaborating with one of the many groups in existence back in 1981. When there is such an alliance of force and when there are reasons to believe that NVATW would not be able to prevent such states from acquiring or developing these weapons, actions that resemble the one undertaken by Israel in 1981 ought to be considered. However, as I have already argued elsewhere (Caron, 2019a), another dimension that ought to be considered is the use of cy- berattacks that can take either the form of NVATW or of VATW. In the case of non-violent actions, we can think of the one directed against the Natanz nuclear facility in Iran by a malware known as 'Stuxnet'. Introduced inadvertently or on purpose by an employee who most likely plugged a contaminated USB drive into the central computers of the facility (which was not connected to the internet), the virus – allegedly created by Israel and/or the US – managed to take control of nuclear centrifuges and caused them to malfunction and self-destruct, while sending contradictory messages to the operators who thought everything was in order. Although Iran never released specific information about the incident, it is estimated that around 1,000 uranium-enriching centrifuges were destroyed, which led to a significant decrease in the country's enrichment efficiency (Broad et al., 2011) and delayed its capacity to potentially develop nuclear weapons by as much as two years (Stiennon, 2015, p. 20). Although very similar to the bombing of the Osirak reactor in terms of destruction, I would argue that the Stuxnet virus was, contrary to the 1981 attack, a clear example of what I am defending as a pre-emptive attack under the lens of credibility rather than imminence. In this case, this cyberattack was not simply launched out of fear that Iran may eventually disrupt the regional balance of power, but rather that the WMD resulting from its nuclear programme could have easily fallen into the hands of stateless terrorist organizations. For instance, it is known that Iran has been providing various kinds of support to terrorist organizations – namely, the Hezbollah and the Hamas. Moreover, alleged links with Al-Qaeda were found in connection with the 1998 attacks against the US embassies in Tanzania and Kenya (Thiessen, 2011), the attack against the USS Cole (Hsu, 2015), and even with the events of 9/11 (The 9/11 Commission Report, pp. 240–241). Additionally, the way in which the Iranian authorities deceived the international community about the nature of its nuclear programme also raised serious questions as to whether it was purely dedicated to civilian purposes. Alongside the discovery of undeclared nuclear facilities – namely, the Natanz complex, a heavy water production plant under construction in Arak, as well as centrifuges that were clandestinely imported in the 1980s – the rhetoric used by then Iranian President Mahmoud Ahmadinejad also contributed to highlighting the possibility that Iran may eventually try to develop a nuclear weapon. Indeed, in 2006, he announced the decision to resume uranium enrichment at Natanz, which led the UN Security Council to adopt Resolution 1696, which was ignored by Iran. When all these elements are taken into account, it is possible to argue that the prospect of a country known for its close ties with terrorist organizations that use indiscriminate means of warfare having a nuclear programme with military dimensions posed potentially serious threats. There were, therefore, solid grounds to resort to this type of alternative to war in order to prevent this programme from ever being completed after other NVATW did not result in altering Iran's policy. At the same time, cyberattacks can also play a constitutive part of VATW. We can think in this regard to cyber-assassinations. Hypothetically, cyber-assassinations would constitute a similar way of dealing with individuals who are involved in terrorist organizations that are posing a credible threat by, for example, remotely gaining control of their pacemakers[65] and delivering a shock in order to kill its users in a way that resembles the assassination of the US Vice-President in the television series *Homeland*, or by taking control of critical functions of their automobiles[66].

As argued in the previous chapter, resorting to VATW against terrorist organizations or states sponsoring such entities has the same foundation as measures imposed against states when the R2P principle is threatened. In both cases, VATW are required and implemented when the threat is deemed credible and not necessarily imminent. This was clearly the case against Iraq in 1991 as the series of containment measures imposed against Saddam Hussein's regime were based not on evidence that he was about to massacre parts of its population but rather on a credible fear rooted in his rhetoric and military capabilities. When this is the case, not acting runs the risk of opening a door to the death of countless innocent people. Although such a situation may eliminate all doubts about the rightness of using force against such a regime, waiting for innocent people to die before we can deem ourselves justified in intervening is morally questionable. This is why, consciously or not, the international community's humanitarian interventions before the initiation of mass human rights violations have largely been justified by the fact that the risk to people's lives was credible and not imminent. In this sense, resorting to VATW against terrorist organizations or states collaborating with them should not be seen as a heretic idea but rather as the application of an already-existing norm to another similar type of threat against civilians.

4.2 The Objectives and Moral Criteria of Alternatives to War

From what has been discussed earlier, VATW are measures that ought to be used solely for the sake of self-defence once NVATW have proved ineffective or when there are reasons to believe that they will not result in preventing a terrorist threat from becoming credible. Accordingly, these actions that are not to be considered acts of war – this is what the next section will discuss – should not be perceived as they were by George Kennan in 1946, namely as a way for a state to

---

65 Some pacemakers have wireless interfaces that allow doctors to adjust their settings at a distance and to share data logs online. As stated in a BBC report, "In 2012, security researcher Barnaby Jack demonstrated an attack using the radio-frequency interface on a heart device. [He] said he was able to launch his attack from a laptop up to 50 ft (15 m) away" (Vallance, 2015).
66 Although this last example may run counter to the discrimination principle, since this malfunction may lead to pedestrians or other drivers being hit.

exert its influence over other states without having to '[reach] for their weapons and shooting it out' (Kennan, 1991, p. 3)[67]. In the realm of the just war tradition, force and violence should not be tools for increasing states' power and dominance over the wannabe hegemons but rather at the service of a higher moral end, that is, saving civilians' lives irrespective of whether they belong to states that are targeted by terrorist organizations or whether they live where terrorists operate from. This is precisely why these measures short of war must be anticipatory by nature as it is necessary to avoid having to face situations like the one George W. Bush experienced on that fateful morning in September 2001 when he was reading a book to schoolchildren. Indeed, despite all we can say about his decision to wage a full-scale war in Afghanistan, we must be honest with ourselves and accept how politics actually works by admitting that he did not really have any other choices at his disposal. After having been struck at the heart by bin Laden and his men, nothing and no one could have calmed the American people down. Just like after the attack on Pearl Harbor, revenge was the only item on the agenda. As we now know, such a reaction – despite being natural and obvious after such incidents – ought to be avoided as a full-scale war against a terrorist organization will have devastating effects, namely political destabilisation of a country and its neighbours as well as the risk of disproportionate civilian casualties. For the sake of avoiding the tragic consequences of war, proactivity is the key and this is why such terrorist attacks must simply not occur at all, and the best way to achieve this goal is by allowing states to have the tools at their disposal to pre-emptively eliminate credible terrorist threats.

This capacity to avoid the possible spill-overs of violence is attributable to the fact that VATW are restricted to specific surgical attacks against legitimate targets, such as strategic infrastructure or individuals, in ways that prevent innocents from harm. In order to make this hope possible, states resorting to VATW need to show a high degree of respect for moral and ethical principles. Indeed, if states have the obligation to safeguard the lives of their citizens, the eradication of the terrorist threat must not come at the expense of other duties, with the most important being the obligation to restrict the use of violence only against those who are legitimate targets. We need to stress the importance of the fact that justifying a more permissive use of political violence does not eliminate in any way the usual ethical constraints statesmen and members of the military ought to re-

---

67 Measures short of war have also been described as 'a term traditionally used to describe all national ways and means available to help policymakers achieve geopolitical objectives *without* crossing the line into major conventional or (since 1945) nuclear confrontation. Kennan lists a range of these, from negotiation to embargo to intimidation, covert subversion, assassination, and the limited use of military force' (Connable et al., 2016, p. 1).

spect. Thus, these additional duties imply that the use of deadly or destructive force will need to be proportional to the nature of the threat and respectful of the discrimination between combatants and non-combatants. In this regard, states will have to consider a wide variety of options ranging from sending elite troops on the ground (whose actions will be limited to a surprise operation with limited usage of weapons) to the use of drones with a much larger scope of destruction. One of the main criteria they will have to consider is the nature of the environment in which the individuals to be targeted are operating. More precisely, irrespective of the seriousness of the threat, the approaches may have to be very different if these individuals are located in the middle of the desert or in a densely populated area.

More precisely, the criteria that ought to be followed when it comes to the use of VATW are very similar to those that have been put forward by the Israeli High Court in a famous decision about the lawfulness of targeted killings (2006). In my view, these are the most important principles to follow:

1. VATW can only be justified against enemies who are, based on reliable, authentic, and confirmed information from multiple sources independent from one another, posing a credible threat to civilians, more specifically that they have at their disposal the means to effectively transform their menace into reality or are actively trying to achieve this goal. Since individuals who will be victims of VATW are deprived of due process of law, a departure from this rule must not be taken lightly;

2. VATW are only permitted if NVATW are thought to be ineffective at eliminating the threat. This means that the resort to VATW must take into account the actual capacity of preventing the terrorist from striking through NVATW. Further, it must not result in disproportionate danger to civilians who might happen to be in the vicinity when the operation will take place or create unnecessary risks to the soldiers who will be deployed[68];

3. The choice of which VATW will be used to eliminate the threat as well as the way it will be used must prevent as much as possible the harm to innocent civilians.

To a very large extent, this logic supporting the resort to VATW is the one used by Israel in its fight against the terrorist groups that have been threatening the lives of its civilians for decades now. Combining reactive measures, such as the use of the Iron Dome weapon system (Caron, 2020), with proactive measures (targeting

---

68 For a discussion on the military's duty of care towards its members, see Caron (2018a, 2018b).

members of terrorist organizations through lethal means of actions), Israel is probably the best example of a state that has privileged this middle ground between soft and hard wars. Of course, this example is far from being a perfect example of proper use of VATW, as many recent operations from the past decades have proved themselves to be largely disproportionate and to have lacked the necessary discrimination between legitimate targets and non-combatants[69]. In a nutshell, this case shows how resorting to VATW is not a panacea and requires a proper respect of these aforementioned rules and those that derive from them.

Obviously, evoking the use of drones as a proper VATW technology that is able to ensure the respect of the discrimination principle may raise some eyebrows in light of the fact that out of the 542 drone strikes authorised by former President Barack Obama that killed an estimated 3,797 individuals throughout the world, 324 of them were civilians (Zenko, 2017). Relatedly, a 2009 report from the Brookings Institute argued that for every terrorist killed at the time by a drone strike in Pakistan, an average of 10 civilians were also killed, thereby making the civilian to combatant ratio an astonishing 10:1 (Byman, 2009). These figures are very paradoxical in light of the fact that this weapon can identify things and individuals on the ground from up to 50,000 feet away. Furthermore, we must also consider the fact that its operators have plenty of time to assess a particular situation before deciding to use lethal force. On paper, a drone seems to be the perfect weapon, but the reality has proved otherwise, leading many to view this technology as immoral (Chamayou, 2015).

If the reality differs from the theory, it is primarily because of the way drones are being used. In this regard, the high number of killings of innocent people by American drones resulted from the use of 'signature strikes', namely attacks against unknown individuals whose behaviours are considered suspicious according to certain patterns-of-life analysis. For instance, individuals seen digging a hole and hiding something on the side of a road will likely be targeted since their behaviours will certainly be interpreted as typical of terrorists planting an improvised explosive device. This operational method is not perfect and has led to many unfortunate mistakes, such as the one reported by Nina Franz:

> In addition to video surveillance, this signature may consist of electronic communications that the CIA is known to collect through technology installed on drones via the National Security Agency (NSA), allowing the US intelligence agencies to gather data indiscriminately on an industrial scale, according to leaked sources. As the publication of the Edward Snowden files in 2013 revealed, the NSA makes use of this data through a machine-learning

---

[69] The operations of May 2021 against members of the Hamas in the Gaza Strip are probably the best examples in this regard.

program called 'SKYNET' (...). The NSA's SKYNET is far from infallible, as indicated by a leaked top-secret slideshow about the program that cites the example of an ostensibly successful identification of a high-level terrorist via pattern-of-life analysis: "The highest scoring selector that travelled to Peshawar and Lahore is PROB AHMED ZAIDAN". The targeted man that the NSA identified as a courier, simultaneously being a member of Al-Qaeda and the Muslim Brotherhood, is, in fact, a well-known journalist working for the Doha-based news network Al Jazeera, whose investigative reporting on terrorist networks had apparently given him the ideal pattern of terrorist activity according to the algorithms of the NSA (Franz, 2017, p. 116).

If that mistake was realised before it was too late, other individuals did not have the same chance. Grégoire Chamayou provides many sad examples in this regard. For instance, on 2nd September 2010, the American authorities announced that they had eliminated a top Taliban leader in Afghanistan with a drone strike. It turned out, however, that the man killed was Zabet Amanullah[70], a campaigning politician. His mistake was the overuse of his cell phone as well as the identity of those he called. This pattern-of-life analysis was deemed suspicious and similar to that of a terrorist (Chamayou, 2015, p. 50). Chamayou also discusses a famous incident that occurred in March 2011:

An American strike decimated a group of men meeting in Datta Khel, Pakistan, on the grounds that "they acted in a manner consistent with AQ [Al-Qaeda] linked militants". The manner of their gathering corresponded to that predefined as resembling terrorist behaviour. But the meeting observed from the skies was actually a traditional assembly, a jirga, convoked to resolve a disagreement in the local community. Seen from the sky, a village meeting looks just like a gathering of militants. Between nineteen and thirty civilians are estimated to have perished in the attack (Chamayou, 2015, p. 50)[71].

If signature strikes have led to the targeting of innocent people, so has the 'double tap' practice by the US military, which consists of targeting the same site in quick succession. This practice has killed numerous first responders who came to the rescue of those who were injured by the initial hit. Of course, this raises serious concerns regarding respect for the rules of warfare. More precisely, as noted in the *Living Under Drones* report:

Not only does the practice put into question the extent to which secondary strikes comply with international humanitarian law (...), but it also potentially violates specific legal pro-

---

70 Alongside nine others in the vicinity.
71 In light of these rules of engagement associated with signature strikes, Chamayou wrote that 'When the CIA sees three guys doing jumping jacks, the agency thinks it's a terrorist training camp' (Chamayou, 2015, p. 49).

tections for medical and humanitarian personnel, and for the wounded. As international law experts have noted, intentional strikes on first responders may constitute war crimes (Cavallero et al., 2012).

However, these examples should not lead us to conclude that drones are fundamentally indiscriminate weapons. Indeed, these aforementioned violations were the result of the inappropriate use of this technology – namely, the propensity to use signature strikes and the double tap practice. The method used by Israel to target some individuals[72] is a clear example of how drones can be used in a way that is inherently designed to be respectful of the moral rules of warfare[73]. The case of Salah Shehadeh is quite enlightening in this regard. Shehadeh was a founding member of Hamas and quickly became one of its leaders, leading to his arrest both by the Israeli and Palestinian authorities between 1988 and 1999. As head of Hamas' operational branch, Shehadeh was directly involved in the planning and execution of 52 attacks following the Second Intifada that led to the deaths of hundreds of Israeli citizens (including many women and children) and the wounding of thousands. For Israel, it became urgent to neutralise the man responsible. At first, the Israeli military tried to arrest him by sending soldiers into the Gaza Strip, but quickly came to the conclusion that such an operation was infeasible because Shehadeh was constantly switching residences, which did not give the Israeli army enough time to prepare a proper plan, and because such a ground operation in a densely populated area would have led to unreasonable risks to Israeli soldiers and civilians. It was only then that the recourse to targeted killing was considered. The decision to kill him was taken on 22nd July 2002. On that fateful day, 14 civilians (including nine children) were killed and more than 70 were wounded.

What is so different between this case and the way Americans have been using drones since the beginning of the war on terror? The difference is that the decision to kill Shehadeh was not taken lightly and that previous attempts had been cancelled on numerous occasions after it was reported that Shehadeh was accompanied by his daughter (Byman, 2009). The decision to strike was taken only after intelligence reported that Shehadeh was alone in a building with no civilians in the vicinity. This highlights a key difference between the Is-

---

72 The question of whether it is morally acceptable to kill an enemy at a great distance and if it contributes to calling into question the legitimacy of killing during warfare will be discussed in a later chapter.

73 Although, as I have said previously, Israel's recent use of targeted killings has also failed in the recent past.

raeli and American rules of engagement: the decision to use targeted killings – whether with drones or conventional aircraft – follows strict procedures that are in place precisely in order to respect the moral rules of warfare. In the special investigatory commission that followed this incident, it became clear that the great majority of senior commanders involved agreed that the operation would had been called off once more had they known that the intelligence report was incorrect; thus, the terrible consequences that resulted from the attack did not stem from their disregard for human life and the rules of warfare. Moreover, when errors occur, tribunals investigate whether criminal charges ought to be filed against ill-advised individuals. In the case of Salah Shehadeh, the results of the investigation – to which the representatives of those who were harmed were invited to participate – were shared through an unclassified report that included information not normally shared for national security reasons (Shelah Shehadeh-Special Investigatory Commission, 2011). At the end of the day, it was discovered that the intelligence failure was the result of objective constraints and not gross negligence, which led the Israeli armed forces to change their procedure regarding targeted killing operations. This comparison between the US and Israel is very useful and shows that the debate around the moral legitimacy of current technologies depends on the way they are used, how the rules of engagement are revised after an unfortunate blunder, and how individuals responsible for negligence are held responsible for their errors. In sum, it is important not to confuse the means with the ends sought by drones. Further, it is important to understand that the problems that derive from their use are not necessarily connected to the weapons themselves but rather with the method of use. From an ethical perspective and the moral duty to use them, this distinction cannot be ignored.

Alongside the obligation to be proportionate and discriminate, measures short of war that are undertaken against a state that is sponsoring terrorism also need to respect another imperative. As said earlier, what the 'war on terror' has taught us over the last two decades is that a full-scale military operation against states harbouring terrorist organizations can lead to political instability with deleterious side effects that will cause the death of thousands of civilians, as in Afghanistan, Iraq, and also Libya after the fall of Gaddafi in 2011. This is why states resorting to these measures should be strictly limited only to preventing a credible terrorist threat. In fact, this ought to be the only goal behind the resort to VATW; they should not become tools for forced regime changes abroad for an obvious reason – the empirical evidence unequivocally suggests that this strategy is more likely to fail than to succeed. Indeed, a forced regime change tends to weaken existing institutions, and the imposition of a new leader from an external power creates grievances against the new regime, weakens its legitimacy, and leads to civil conflicts that can degenerate into a full-fledged civil war (Ferwerda & Miller, 2014). This strategy may also destabilise a whole region, leading to civil unrest and the death of thousands of civilians as was the case in the Middle East after the end of Saddam Hussein's reign of tyranny in Iraq. This is why VATW ought to be used simply for the sake of self-defence, meaning that once a threat ceases to be credible, no further actions should be taken. At the end of the day, the lives of our citizens are as valuable as the lives of those who have the misfortune of living where terrorists operate, and states that are being threatened by this menace need to act accordingly.

4.3 A Revival of 19th-Century Measures Short of War

The limited nature of violence associated with VATW helps correct an unwanted understanding of force and the use of kinetic actions as being akin to acts of war. Correcting this false impression requires us to define what constitutes an act of war. Contrary to the general assumption, if 'war is an act of force' according to von Clausewitz (p. 76)[74], it would, nonetheless, be a mistake to conclude that all resorts to force and violence are acts of war (Gartzke, 2013, p. 54). Furthermore, many violations of sovereignty are not treated by states as wartime actions or as actions that ought to trigger a right of self-defence. As Walzer says about the enforcement of no-fly zones, embargoes, or the bombing of aircraft installations, it is common sense to perceive them very differently from acts of war (2006, p. xiv). Relatedly, we could add to the list isolated border clashes between rogue soldiers or naval incidents between two vessels after one has entered into another state's territorial waters. What distinguishes these acts of violence from war is their intensity. In other words, it is recognised that a certain threshold has to be crossed before an action ought to be treated as a *casus belli*. There are obviously many disagreements about what constitutes that threshold. One mistake would be to use David J. Singer and Mel Small's definition of war, which requires 'at least 1000 battle deaths' (1972). Obviously, such an understanding of war is problematic and raises the question of why the threshold is set at 1,000 deaths and not 500 or even 995.

This red line would, according to Brian Orend, be the resort to an intentional and heavy quantum of force (or its imminent use) that is preceded by a signifi-

---

[74] The Resolution on the Definition of Aggression also defines war as 'the used of armed force by a State against the sovereignty, territorial integrity or political independence of another State (...)'.

cant mobilisation of military personnel and the deployment of military equipment on a large scale (2006, p. 2–3). According to this definition, the 2003 invasion of Iraq was clearly an act of war. Of course, it is possible to quarrel about the nature of the intensity of force that is required before reaching the threshold that allows for the differentiation between an act of war and an action that falls short of being considered as such. According to Jessica Wolfendale, this intensity ought to 'be measured by the level of disruption caused by a conflict to those living in the arena of conflict (combatants and civilians), including the impact on their physical safety; access to basic goods such as food, water, warmth, and shelter; and the functioning of basic civilian infrastructure'. In other words, 'a conflict meets the criterion of intensity when it becomes so disruptive that the ability of civilians to meet their basic needs is seriously threatened, and the local authorities are unable to effectively control the conflict and protect civilians and civilian infrastructure from harm' (Wolfendale, 2017, p. 21). VATW are far from meeting that threshold when they are used in accordance with the guidelines previously described.

This is why it is problematic to establish as it is the case with Michael Gross and Tamar Meisels a dichotomy between NVATW defined as 'soft war' measures[75] and those involving violence as being constitutive of 'hard war' measures. With such a distinction, actions that are not considered 'soft' become *de facto* akin to 'the stuff old-fashioned wars are made of' (2017, p. 1). As VATW have nothing to do with what 'old-fashioned wars' were about and because there are reasons to argue that they are not acts of war, I find it rather unfair to present them as equivalent to 'mass killing fields [and] wholesale deaths in battle (...)' (2017, p. 1) as if there were only one form of political violence. If it were to be the case, the death of one soldier in a border clash between combatants would then be qualified as an act of war. This is why NVATW and VATW should simply be called measures short of war as they bear a lot of similarities with the sorts of actions that were understood as such in the 19th century and during the first half of the 20th century. These measures, accepted by the international community as not being akin to acts of war, were mainly seen as measures of law enforcement and were limited in terms of scope and aims. Political interventions within the affairs of other states fit this definition. Stephen C. Neff reminds us in his seminal book on the evolution of conflicts (see particularly chapter 6, 2005) that the European powers that pre-

---

75 For them, the notion of soft war refers to non-kinetic actions that fall short of armed attacks, such as 'bytes, boycotts, propaganda, non-violent resistance, and even kidnapping' rather than 'bullets and bombs' (2017, p. xv). In this sense, their understanding of such actions is very close to Pattison's view of NVATW.

vailed against Napoleon in 1815 were primarily motivated to prevent any kind of future conflicts on the continent and gave themselves – France joined them in 1818 after it was safely back under the rule of Louis XVIII – the responsibility of maintaining the status quo and the balance of power. This is why Austria intervened in Naples and Sardinia in the 1820s in order to restore peace and stability after domestic events had led to the destabilisation of these kingdoms. It is also following this logic that France intervened in Northern Spain and restored the authority of King Ferdinand VII after insurgents had taken control of that part of the country or why a naval blockade was imposed against Greece in 1886 after it had launched an offensive against Bulgaria. Similar actions were also undertaken for the sake of what would later be known as 'humanitarian intervention' in order to prevent the massacre of a civilian population. This was the case in 1860 when France sent troops to Lebanon, when communal violence posed a threat to the lives of civilians.

Reprisals were also thought to be measures short of war that did not amount to acts of war; rather, they were seen as forceful means for the sake of pursuing a just cause. In fact, the 1907 Hague Convention on the Opening of Hostilities made it clear that reprisals were not akin to acts of war. These actions were seen as justified only after a state that had been the victim of an injustice had expressed its grievances against the state responsible and asked for reparatory measures. If the solution was not satisfactory, the former state was allowed to employ forceful actions that were deemed proportionate to the gravity of the offence; actions that had to cease immediately after a satisfactory solution had been obtained. A good example of reprisal was the French seizure of the Turkish port of Mytilene in 1901. Neff writes the following:

> The purpose was to induce Turkey to provide satisfaction to France for a number of alleged infractions of international law to the detriment of French nationals, which France carefully identified in a diplomatic note. There was no violence or destruction. Moreover, the action was successful in inducing Turkey to reach a settlement of the dispute with France, after which France duly evacuated the captured area. It was observed, apparently without irony, that the incident was "a truly ideal reprisal", involving no loss of life, no infringement of the interests of the third parties and a wholly satisfactory outcome (for France, that is) (Neff, 2005, p. 228).

Emergency actions employed by a state in order to defend itself from what it considered an imminent danger were also deemed measures short of war. This was the case in 1807 after the British destroyed part of the Danish fleet in the port of Copenhagen and took control of the rest despite the fact that the two countries were not at war. The reason was simply to prevent Napoleon's troops, quickly approaching the Danish capital, from taking control of the fleet and using it against the Royal Navy. A more contemporary example would be the destruction of the French fleet at Mers-el-Kébir in 1940 after Marshall Pétain had signed the armistice with Germany. Churchill and the British Cabinet feared that the powerful and modern battleships stationed in the harbour would fall into German hands and then be used by Hitler in his openly stated intention to invade the British Isles. Following an ultimatum that was left unanswered by the French, orders were given to British Admiral James Somerville to destroy the fleet. The attack, which was described by Churchill as 'the most hateful decision, the most unnatural and painful in which [he had] ever been concerned', led to the death of more than a thousand French sailors.

The rescue of nationals in peril abroad as well as punitive expeditions were also considered measures short of war. The British government resorted to the former action in 1868 against Ethiopia after its Emperor decided to take hostage a consular official and a government envoy. After an ultimatum had been rejected by the Ethiopian Head of State, the British Army launched a successful military expedition that stopped after the hostages were freed. For their part, punitive expeditions were also tolerated against individuals or groups who had committed wrongdoings against a foreign state. This was the case when the American military entered Mexico in 1916–1917 to capture Pancho Villa after he had killed American citizens taken from a train in Northern Mexico and after he had burned down a city in New Mexico, which also resulted in the death of 19 American citizens.

As argued by von Clausewitz, because of their aim, all these measures were not considered akin to acts of war mainly because 'the hostile spirit of a true war' was lacking. On the contrary, they were perceived positively because measures short of war did not possess an *animus belligerendi* on the part of the state using them. They were used not to promote mere selfish national interests but for noble intentions such as preventing instability, saving lives, self-defence, or punishing individuals who had committed crimes. Moreover, the limited impact of these measures and their low intensity allowed statesmen and lawmakers to distinguish them from acts of war. In many ways, NVATW and VATW share the same fundamental goal as that underlying the application of measures short of war in the 19th century and the early decades of the 20th century, namely to prevent the outbreak of a full-scale war.

The post-1945 order, however, banned the resort to violent measures short of war as the drafters of the UN Charter sought to forbid all forms of armed violence irrespective of their degree of force (except for the sake of self-defence or when approved by the Security Council), thereby eliminating that distinction between war itself and alternatives to war that involved the use of force (Neff, 2005, p. 314). However, NVATW were not affected. Indeed, the scope of force forbidden in Article 2(4) of the UN Charter has been generally interpreted as not including non-military type of force, such as economic sanctions or the suspension of a treaty obligation (Elagab, 1988; Henderson, 1986)[76]. As I have argued so far, this black and white view of political violence is problematic as it leaves states vulnerable to certain types of threats, especially when it comes to terrorism. Remaining faithful to that vision would mean that when NVATW prove ineffective against this danger, states only face the choices of inaction (which may imply letting their citizens being killed, an option that will most likely trigger a desire for revenge and lead to a full-scale war) or by taking proactive forceful measures (which may be seen as acts of war). This is why we need to agree on the fact that force encompasses a spectrum of degrees of violence that can be akin to acts of war only when a certain threshold is crossed, a threshold that VATW as I have defined them does not meet. When this is not the case, everything becomes either war or peace (Lupton & Morkevicius, 2019, p. 44).

Whether we like it or not, customary international law makes this distinction when it comes to VATW understood in terms of reprisals and it may just be a matter of time before anticipatory VATW follow the same path. Indeed, over the last 40 years, many violent actions that followed an attack were deemed to fall short of acts of war despite being technically unlawful (Magenis, 2002; Brennan, 1999). In this case, their main objective was not self-defence – as is the case with preemptive measures – but rather the desire to punish an aggressor and deter future attacks. This was the case in 1986 when the US launched air strikes against half a dozen military installations in Tripoli in response to a terrorist attack in a West Berlin nightclub commonly frequented by American soldiers. The Libyan government was accused of sponsoring the attack, which caused the death of two American servicemen and injuries to 79. Furthermore, during Operation Earnest Will from 1987 until 1988, the US engaged in retaliatory actions against Iran after tankers and vessels were attacked by the Iranian military in the Strait of Ormuz. More precisely, Iranian oil platforms were bombed and destroyed by the US Navy. In 1993, after a plot to assassinate former President Bush during his visit to Kuwait was thwarted, President Clinton nonetheless ordered the firing of 23 cruise missiles against Iraqi military targets. Finally, following the attacks against the US embassies in Tanzania and Kenya in 1998, the US launched 79 cruise missiles

---

76 In fact, a subcommittee rejected in 1945 by a vote of 26–2 an amendment to Article 2(4) made by Brazil that would have included economic sanctions as a form of forbidden coercion.

against targets linked with Al-Qaeda in Afghanistan and Sudan after it was determined that the organization was responsible for these acts of terror[77].

As VATW, reprisals are proper measures short of war that ought to be employed by states following a terrorist attack rather than full-blown wars. Indeed, as the last 20 years have shown us, war is not only a disproportionate measure against these groups that comes with the previously evoked consequences but also a hugely ineffective one. Contrary to a state that can be defeated and against whom peace can be achieved following an armistice or a formal peace treaty, terrorism can only be prevented with the appropriate anticipatory measures that have been discussed or contained at best after these groups have already struck[78]. The large deployment of troops can do very little against an idea that will end up inspiring lone wolves or small cells all around the world and who will strike devastating and murderous blows at civilians who are calmly enjoying an evening walk on a boardwalk or a drink on a terrace, or simply reading a book on the subway on their way to work. This Sisyphean strategy is doomed to bring endless wars and destruction as well as engendering a situation that will paradoxically lead to what states are supposed to prevent, that is, the death of tens of thousands of innocent civilians. This is why these sorts of VATW that are accepted forms of violence – even if they are still illegal according to international law – ought to be considered the right form of retaliation following a terrorist attack, rather than war.

Contrary to anticipatory VATW, justifying these violent reprisal measures that fall short of war may be easier because they result from an attack that has already occurred; in such a situation, even those who have a restrictive view of self-defence may be convinced. It is, of course, another story when an attack has not occurred and does not meet the conventional criteria of pre-emp-

---

77 John Yoo (2003) presents the 1986 bombing of Tripoli and the 1998 retaliatory actions against Al-Qaeda in Afghanistan and Sudan as being a form of anticipatory self-defence. Since they were not condemned by the UN Security Council, he believes these precedents justified the 2003 invasion of Iraq. His conclusion derives from the fact that he is not making an essential distinction between anticipatory and reprisal measures. The fact that the latter tend to be accepted by the international community does not necessarily mean that it is also the case for the former measures.

78 This unique nature of the war against terrorism was rightly acknowledged by President Bush during his 20th September 2001 joint address to Congress in which he warned Americans that 'This war will not be like the war against Iraq a decade ago, with a decisive liberation of territory and a swift conclusion'. At that time, his appraisal of how terrorists had to be fought was appropriate as well. Instead of thinking of conventional means of warfare, he rather evoked '(...) a lengthy campaign, unlike any other we have ever seen [which] may include dramatic strikes, visible on TV, and covert operations, secret even in success'.

tion. However, sticking to such a cramped definition comes with an obvious risk; as they do not have the capacity to protect their civilians against terrorism, states will be tempted to launch preventive wars as was the case against Saddam Hussein and Iraq in 2003. In light of the current deficiencies, alternatives such as the one defended so far need to be found. Not acting in this regard means that every single resort to anticipatory VATW will simply contribute to eroding the value and relevance of international law in favour of subjective and questionable assessments of the terrorist menace that may well come with disproportionate and indiscriminatory measures. At the end of the day, as argued by Oscar Schachter, acting outside the law – even for legitimate reasons – runs the risk of losing the name of law and creating global cynicism (1984). This ought to be avoided at all costs, which is why anticipatory VATW as a response to contemporary terrorism need to become an integral part of international law.

However, the fact that VATW should not be understood as acts of war poses a significant ethical problem, namely how we can justify the resort to lethal force against individuals. Indeed, killing is thought to be justifiable only when there is an ongoing war between two entities, which is indeed problematic if we are to consider lethal actions undertaken against terrorist organizations or the states sponsoring them as falling outside that premise. If these actions fall short of being constitutive of war, then it implies the necessity of rethinking the legitimacy of killing outside of this paradigm. This is what the next chapter will study.

Chapter 5:
Justifying Lethal Alternatives to War

According to the theory of natural law, which is one of the foundations of modern societies, all human beings have been endowed by their Creator with equal natural and inalienable rights, the most important being the right to life. Accordingly, depriving someone of that right should not be taken lightly and this is why just war theorists have spent a great deal of time and effort justifying the violation of this right. This chapter will examine how this justification can be applied to VATW.

5.1 The Justification of Killing During Wartime

Although fundamental, the right to life is not absolute. Apart from the highly debated question of capital punishment, individuals can indeed make themselves liable to just harm when they threaten to deprive others of their right to life. This is the whole basis of the logic of self-defence. Indeed, when individuals face a lethal threat, they are justified in taking all necessary and proportionate measures to defend themselves. Therefore, killing an individual who is trying to steal your car or wallet is unjustified owing to the disproportionality of the response. This is because the offender is not posing a threat to your life, and resorting to lethal force is not necessary to avert the violation of your property despite the unlawful nature of his actions. This conceptualisation of self-defence is at the heart of the justification of killing during wartime. Indeed, for Michael Walzer, all combatants are thought to pose a potential deadly threat to their foes once a war has begun. The reason is that 'soldiers as a class are set apart from the world of peaceful activity; they are trained to fight, provided with weapons, required to fight on command. No doubt, they do not always fight; nor is war their personal enterprise. But it is the enterprise of their class (...)' (Walzer, 2006, p. 144). This explains the essence of the principle of discrimination and why soldiers should not deliberately harm civilians. If combatants pose a potential lethal threat to other combatants, this is simply not the case with individuals who are not soldiers. As summarised by Jeff McMahan:

> Those who retain their immunity to attack are therefore those who are not threatening. In the context of war, the innocent are those who do not contribute to the prosecution of the war – that is, non-combatants. The non-innocent are those who pose a threat to others – that is, combatants. They lose their immunity and are liable to attack (McMahan, 2006, p. 24).

https://doi.org/10.1515/9783110729894-007

The loss of immunity will, therefore, only affect soldiers, irrespective of the side they fight on, because they can all potentially harm other combatants by joining their country's armed forces. Known as the 'symmetry thesis', this refers to the moral equality of combatants. In return, since non-combatants are by definition unarmed and do not pose a threat to anyone, killing them would not be a matter of self-defence but murder. This category includes civilians and members of the military whose duties make them harmless – like chaplains[79] – or who are prisoners of war, are surrendering, or who are defenceless because of unconsciousness or wounds[80]. When soldiers are in combat on the battlefield, resorting to lethal violence is thought to meet the criterion of necessity as shooting first at an individual who is armed and trained to obey at any time an order to kill you is the only way to prevent him from harming you. Indeed, it would be inconceivable for, say, a sniper to scream at the enemy soldier he has in his sight that he should surrender or be shot. Not only would it allow this enemy combatant to escape from the threat and fight another day but it would also allow his comrades to know where the sniper is hiding and take the proper measures to take him out. In return, soldiers have the obligation to use proportionate actions when they are in such a situation as a way to prevent as much as possible any collateral damage that may result in the harm of individuals who have not lost their immunity against harm.

This unique status of soldiers also opens the door to the targeting of those who have been labelled 'naked soldiers', that is, members of the military who are engaged in actions that are not battlefield activities and who do not pose a direct threat (Deakin, 2014, p. 321). Although Walzer admits that shooting an enemy combatant who is taking a bath in a river behind enemy lines might be psychologically troublesome for many individuals who could feel that because of his vulnerable condition the soldier in question is no longer an enemy but rather a 'normal man', he still believes that this person remains a legitimate target. For Walzer, the logic behind such a conclusion lies in the fact that the naked soldier is not similar to an enemy combatant who has surrendered or is wounded and unable to continue fighting. In such cases, these soldiers are regaining their status as non-combatants and should, accordingly, be treated with humanity and respect. Not doing so would be a war crime. However, as von Clausewitz has argued, all soldiers, even the naked ones, have abandoned their status of 'normal men' by joining the ranks of the military and by submitting themselves

---

79 Protocol 1, 8th June 1977, Article 43.2 states that chaplains are non-combatants and that they do not have the right to participate directly in hostilities.

80 Provided that they are not showing any sign of hostility or trying to escape.

to its martial virtue (von Clausewitz, 1976, p. 144). According to this logic, the naked soldier still remains in a position to harm other soldiers in the future. It is in this sense that Stephen Deakin wrote that 'like a tank, artillery piece, or a military aeroplane, the naked soldier is a weapon of war. Destroying tanks, guns and the like is a legitimate and desirable activity in war whether they are in use or not at the time, and the same is true of naked soldiers' (2014: 329). Even if the naked soldier or unused military equipment do not pose an imminent danger, they may, nonetheless, pose a credible threat because they may resort to or be used in lethal purposes against other individuals, which justifies their killing or destruction.

How does this theory apply to terrorists? Let us first divide the resort to lethal VATW into cases of anticipatory self-defence and reprisals following a perpetrated attack. In both cases, while this may imply the same measures, namely the targeted assassination of members of terrorist organizations, the biggest difference between these two situations is the fact that only in the latter case is deadly violence being used after terrorists have struck the first blow. In case of a pre-emptive assassination against a credible threat, we are the ones who are first resorting to violence. But for the time being, let us discuss lethal violence in cases of reprisals.

We can say that war is triggered once members of terrorist organizations have perpetrated an attack against a state and are, consequently, exposed to a reciprocal risk of being harmed by their enemies. As they have shown their willingness and capacity to resort to violence, it is legitimate to expect that they might repeat this action if given the chance. Thereby, as a matter of self-defence, it is considered necessary to kill them before they have a chance to kill us. Of course, resorting to this action ought to be done in a proportional way, which means following strict rules of engagement that will minimise the harm to non-combatants. However, as argued previously, this VATW will be justified insofar as NVATW have proved ineffective or too dangerous. For instance, alternatives ought to be entertained, such as the arrest of those responsible so they can stand trial, or their abduction, similar to what happened to Adolf Eichmann. Many factors will determine the feasibility of these measures, such as the degree of collaboration or the absence of collaboration on the part of the state from which the terrorist organization is operating or the overall danger to troops on the ground as well as any civilians in the vicinity.

From this perspective, alongside the already discussed case of Salah Shehadeh, many other examples can show that targeted killing ought to be preferred over any other NVATW[81]. The case of Jacques Mesrine, a French criminal who be-

---

81 As G.E.M. Anscombe wrote, 'the stabler the society, the rarer it will be for the sovereign au-

came famous in the 1970s for numerous bank robberies, kidnappings, and prison escapes, is particularly enlightening in this regard[82]. In November 1979, he was shot dead by police officers while his car was waiting at a red light in downtown Paris. What led the police to kill him without warning was one of Mesrine's declarations in an interview to *Paris Match* magazine in which he said: 'I will never surrender. The thing about receiving cops with champagne is over[83]. Now, it is war. I will shoot at them and if civilians are unfortunately victims of my bullets, well so be it' (Francesoir). By saying this, Mesrine made clear that any attempts to arrest him would have ended in a bloodbath, which allowed police officers to shoot him by surprise while he was not in a position to return fire[84]. This killing cannot be compared in any way with manhunting, even if Mesrine's body was riddled by 19 bullets at close range while being strapped to his seat. In light of his intentions, Mesrine was by no means comparable to an elk quietly drinking water from a river and shot from a distance by a skilled hunter. Despite being caught when he was momentarily defenceless, he nonetheless remained at the time of his death a threat to every French citizen's right to life. Therefore, the state had the responsibility to stop him from doing what he had promised to do in this fateful interview.

The same can be said with regard to Osama bin Laden, whose death cannot be attributed in any way to a simple desire for revenge – which is, of course, morally distinct from self-defence – for the murderous deeds he had previously performed. The action taken against him in his Abbottabad compound was a case of self-defence, and his unwillingness to surrender to the Navy SEALs made his killing a matter of necessity. As written by Bradley Jay Strawser:

[As a matter of self-defence]
Importantly, intelligence gathered at his compound after his death has confirmed that UBL was continuing to recruit suicide bombers and meticulously plan and orchestrate further attacks against nonliable people up until the time of his death. His involvement in al-Qaeda operations was shown to be much greater and more active than many intelligence specialists had previously thought he still was at the time he was killed. Indeed, at the time of his death, he was actively involved in many of the major decisions, logistics, planning, execution, command, and coordination of several on-going al-Qaeda terrorist plots to

---

thority to have to do anything but apprehend its internal enemy and have him tried; but even in the stablest society there are occasions when the authority has to fight its internal enemy to the point of killing (...)' (1981, p. 53).

82 A two-part film entitled *Mesrine: Killer instinct* and *Mesrine: Public Enemy Number One* featuring Vincent Cassel was produced in 2008.

83 He had become famous for receiving police officers with champagne when he was arrested in 1973.

84 Two grenades and a bag full of loaded weapons were later found in his car.

kill nonliable people. Hence, (...) he was culpable for posing an on-going and present threat to innocent lives at the time of his death (Strawser, 2014, p. 16).

[As a matter of necessity]
The SEALs continued up through the house, where they finally came upon UBL [Usama bin Laden] himself at the top of the 3rd-floor staircase. There is significant debate over what happened next. Some reports claim that upon seeing the SEALs UBL did not surrender and instead quickly moved into the adjacent another room. (Some have referred to this action as a "tactical retreat" that could be interpreted as a continuance of the firefight; others have argued that it was a non-hostile act and should not have been so interpreted. In either case, it was not an act of surrender as recognized by traditional standards of the law of armed conflict.) The Owen account claims that shots were fired at UBL in the hallway that may have hit him before he disappeared into the adjacent room. Other accounts differ on this point. In either case, the SEALs followed him into the room, assuming (it is claimed by several reports) that he was going for a weapon or some other form of offensive capability (such as a bomb). Upon entering the room, the SEALs saw UBL with two women and some children in the corner. The women and children, who appeared to be of no threat, were not harmed. UBL was shot twice, fatally. Again, it is unclear and contentious whether the shots that killed him occurred in the hallway before we went back into the room, or after the SEALs entered the room. Two loaded weapons were found in the room: an AK-47 assault rifle and a 9 millimetre semi-automatic Makarov (Strawser, 2014, pp. 19–20).

It can be argued that resorting to lethal VATW against members of terrorist organizations after they have perpetrated an attack can be defended based on the conventional justification of killing during wartime. When they are choosing to resort to this sort of attack (such as 9/11 or the 2004 and 2005 bombings in Madrid and London, respectively), their actions can legitimately be labelled acts of war. Indeed, they meet Brian Orend and Jessica Wolfendale's understanding of an act of war: an intentional and heavy quantum of force by those involved as well as a significant mobilisation of combatants striking on the largest possible scale. Moreover, for those who remember or had a first-hand experience of these events, it was obvious that the authorities were unable to protect civilians and civilian infrastructure from harm. Moreover, these attacks have had numerous consequences for people's lives, with the most damaging being their psychological impact when we fly or at the sight of a truck moving strangely in a crowded area.

In this perspective, the reader has most likely understood that the definition of an act of war is relative to the strength of the entities that are resorting to force. Indeed, the mobilisation and effect of these aforementioned cases are far from being comparable by any stretch of the imagination to Germany's invasion of Poland on 1st September 1939 or of Kuwait by Saddam Hussein's armed forces in August 1990. What is important to keep in mind is that all these violent actions and their consequences were the result of the maximal capacities to cause harm that these entities had at their disposal. In return, when states with hundreds of bombers or fighter jets and tens of thousands of servicemen at their disposal choose to send a squadron of Navy SEALs or use one drone for a one-time only surgical strike against an identified terrorist, this act of force falls short of an act of war. This is because even though the number of individuals involved is higher than those in a terrorist attack[85], the quantum of force is exponentially lower than that used by terrorists, who go all-in with the weapons at their disposal. Alongside the scale and effects of damage caused by a resort to violence, this is an illustration of the intentions that animate the actions used by an entity when it is resorting to force and an additional way to distinguish between an act of war and a measure short of war. From a proportional perspective, the 9/11 attacks as well as those in public transport in Madrid and London are similar to the Nazi invasion of Poland or the 2003 invasion of Iraq. As they are acts of war[86], they create a situation of self-defence that can justify lethal VATW from a moral perspective.

Following what has been said so far, justifying the resort to lethal VATW against terrorist organizations under the lens of reprisals is not problematic and follows the conventional logic of killing during wartime. On the contrary, justifying similar measures under the logic of pre-emption is more arduous as these lethal actions are not thought to be performed when states are not at war.

5.2 The Justification of Killing in the Absence of War

As said previously, the sorts of VATW that have been defended so far fall short of being considered acts of war. When it comes to VATW under the umbrella of reprisal measures, this situation does not affect the morality of killing as retaliating in this fashion against those responsible for an unjust harm to civilians is a matter of acceptable self-defence as long as our reaction is proportionate. But this situation makes it more problematic in cases of pre-emptive VATW because measures undertaken in such circumstances are not resorted to when there is an open and ongoing conflict between two foes. In fact, in these particular situations, since states are the ones striking the first blow, do the targets have a

---

85 The 2005 attacks in London were performed by 'only' four individuals, while 19 terrorists took an active part in the 9/11 attacks.

86 Although they are acts of war, it does not necessarily mean that states being targeted by these groups should wage a full-blown war against them and – when it is the case – the state harbouring them. Even if they have a theoretical right to do so, the most efficient and ethical way to fight these groups is probably through VATW for the reasons evoked in this book.

right of self-defence and retaliation? If this were to be the case, pre-emptive VATW would simply be like opening a Pandora's box of further violence that would be deemed legitimate by those pursuing them. The prospect of resorting to such pre-emptive VATW against members of terrorist organizations who are posing a credible threat also comes with another inherent ethical problem that does not affect measures undertaken in cases of reprisals, namely the fact that the risk of death is no longer reciprocal, which transforms killing into a form of manhunting that has nothing to do with self-defence. Such an assumption is, in my opinion, unfounded.

It is true that resorting to lethal violence against an individual is in itself a crime that deserves retribution and reparation. It also justifies that individual's right to use deadly force as a way to defend himself against the aggressor, which is how wartime killing has been justified. In this case, because the risk of death is reciprocal, the necessity of using lethal force exists. This is why killing during wartime is thought to be analogous to a duel (von Clausewitz, 1976, p. 13). On the contrary, killing an unsuspecting enemy seems fundamentally different and has more to do with murder than with self-defence. Moreover, since VATW tend to be undertaken with drones, with their operators having the capacity to strike in safety by being sometimes thousands of kilometres away from their targets, the reciprocity of death has also disappeared from the equation. Such a situation has been termed by Paul W. Kahn, who has no moral objections when combatants are in a relationship of mutual risk, as the 'paradigm of riskless warfare' (2002). However, the issue becomes more problematic when an army is able to destroy its enemies without any risk to its members' lives (Kahn, 2002: 3) and, as such, the paradigm of riskless warfare has more to do with a condition of manhunting and in which the enemy has no way to escape his gruesome fate.

However, benefitting from a right of self-defence is only possible insofar as the victim of lethal aggression has not done anything to lose his immunity against being harmed. This is not the case with the terrorist who, despite not having perpetrated any harm, is still a credible unjust threat to innocent people because he will perform murderous deeds if given the chance. But the whole ethical problem we have to face is precisely the fact that is not like Jacques Mesrine, who was on his way to the closest bank with a fully loaded 9 mm around his waist and an AK47 lying on the passenger seat or Richard Reid, aka the 'Shoe Bomber', who wore an explosive device in his shoes during a flight from Paris to Miami in December 2001. The threat he poses is not imminent, as was clearly the case in the Mesrine and Reid examples. Intercepting Mesrine while he was on his way to the bank (with lethal means if he were to resist with the arsenal in his car) or subduing Reid while he was attempting to light up the fuse in his shoes are without a doubt legitimate pre-emptive actions to prevent an imminent threat from becoming real. In such cases, the actions of these individuals made their intentions clear beyond any reasonable doubt. In the case of terrorists posing a credible threat, they have not yet – like Mesrine and Reid – got in the car with their guns in the direction of the bank or got on a plane to blow it up. Their threat is not yet real. In this sense, we need to wonder if we can deprive people in this position of their right to life. Resorting to NVATW, such as arresting them so they can face trial, is not a morally problematic solution. Indeed, following the far too many shootings in schools and other public places in the US, many individuals who had made threats to gun down their classmates or co-workers were arrested and prosecuted (Almasy, 2019). For instance, according to the Canadian criminal code, individuals who make direct or indirect death threats online face up to five years in prison even if 'they did not really mean it'. If prison can be considered a reasonable sanction, nobody would dare suggest that they ought to be killed before they act since their threat is purely a matter of intentions and not of active preparations. So, the question remains: how is it possible to justify pre-emptive VATW against terrorists?

I would argue that the very often inaccessible nature of terrorists combined with the scale and effects of their potential attacks justifies such measures. Indeed, arresting people suspected of posing a threat is an option states can entertain as a measure of law enforcement because these people are enjoying the monopoly of violence on state territory. In cases when it is believed that arresting them will not result in carnage (as it was ultimately the case with Mesrine), necessity dictates that non-violent alternatives be used against them. An operation primarily planned to arrest a potential criminal will only rightfully evolve into a killing enterprise if the targeted person threatens to use or actually uses lethal means against the police or innocent people in the vicinity.

Things are unfortunately different in cases when this method cannot be entertained at all. When terrorists posing a credible threat to the lives of innocent people are elusive and uncatchable, we need to objectively wonder if there are any reasonable alternatives other than lethal VATW. Obviously, these measures ought to be a last resort as requested by the criteria of necessity and proportionality, but this is a viable solution only when alternatives are possible. And this is the Cornelian dilemma that states being threatened are facing: the choice is either between inaction and hoping that the menace is yet another of the empty threats we have become accustomed to in recent years owing to social media, or taking them seriously. If the latter option prevails, because these are serious and credible life-threatening menaces that cannot be deterred non-violently, the resort to lethal force is the only proportionate and necessary measure available.

If this proposition is accepted, it does not mean that those being subjected to these lethal VATW ought to claim in return a right of self-defence. In fact, defensive force is not permissible when the individual who is targeted by lethal means – because it is necessary and proportional – has engaged in an action that renders an attack against him necessary. For instance, as argued by Jeff McMahan (2004), an individual who is chasing you at night with a Rambo-style knife in his hands screaming that he will stab you to death when you are calmly walking your dog in a park cannot claim to have had a right of self-defence if he ends up killing you after you have chosen to defend yourself with lethal force (let us say, you jump on him with a big rock with which you are attempting to hit his head violently). Similarly, an armed bank robber cannot claim to have a right of self-defence after killing a security guard who was trying to reach for his weapon while the criminal was threatening to shoot the cashiers if they refused to hand over the money. By defending themselves against this unjust harm against their lives, the man walking his dog or the security guard do not become as non-innocent as their attackers and, as a consequence, their self-defence does not trigger a similar right on the part of those against whom it is used. This is why McMahan writes accurately that:

> Most find it impossible to believe that, by unjustifiably attacking you and thereby making it justifiable for you to engage in self-defense, your attacker can create the conditions in which it becomes permissible for him to attack you. Most of us believe that, in these circumstances, your attacker has no right not to be attacked by you, that your attack would not wrong him in any way, and that he therefore has no right of self-defense against your justified, defensive attack (McMahan, 2004, p. 699).

Walzer famously refused to apply this logic to soldiers participating in unjust wars of aggression by arguing that it is not comparable to individuals who are involved in criminal activities. If being involved in a criminal activity is a matter of choice, it is different when it comes to fighting a war. According to him, war sometimes implies an explicit obligation on the part of soldiers – for instance, when military service is mandatory or when the draft is imposed – or implicit pressures such as patriotism (2006, p. 28)[87]. Moreover, we should add to his explanation the impact of the civil-military divide in liberal societies where the decision to go to war is solely a political matter in which members of the military do not have a say (Caron, 2019a, p. 12; Huntington, 1957). Accordingly, even if crimes against peace constitute an aggression, soldiers are not the ones who

---

[87] McMahan adds: 'Because those who become combatants are subject to a variety of forces that compel their will – manipulation, deception, coercion, their own sense of the moral authority of the government that commands them to fight, uncertainty about the conditions of justice in the resort to war, and so on – they cannot be held responsible for merely participating in an unjust war' (2004, pp. 699–700).

are ordering them and their decision to be involved in them is beyond their control.

If this conceptualisation may apply to soldiers, it simply does not to members of terrorist organizations. Indeed, members of terrorist organizations are not explicitly compelled to join these groups, while patriotism cannot be compared in any way with the activities pursued by terrorist groups, as the love for a country and the willingness to defend its freedom and institutions bears significant moral differences from the desire to kill people. If the former love is morally neutral, the latter is the quintessential expression of evil. Furthermore, contrary to conventional soldiers, members of terrorist organizations directly participate in the design, planning, and decision to strike against their targets as was the case with those involved in 9/11. This means that they cannot claim a right of self-defence when they are targeted through pre-emptive VATW as they are not soldiers or ordinary civilians, but rather 'civilian criminals' whose decision to participate in an unjust endeavour is not made in the absence of absolute freedom (unlike soldiers). Just like a thief committing an armed bank robbery, they are themselves the root cause of the unjust situation we are trying to prevent and, consequently, do not have any right to defend themselves or retaliate[88].

With this in mind, the second criticism against the legitimacy of killing individuals remotely (with the use of drones being a privileged VATW) – the fact that it is akin to manhunting – is weakened (Chamayou, 2015). Indeed, terrorists who pose a credible threat to civilians are in a similar position to the naked soldier who is taking a bath. The fact that at the moment of being targeted he is not posing a threat to the operator does not mean that he will forever remain in that position. If he is left in peace and allowed to continue planning his murderous deeds, he will eventually pose a life-threatening risk to others. This being said, some may be tempted to argue that there is a major difference between a sniper with a naked soldier in his sight and a drone operator as the latter is not on the battlefield but thousands of miles way. In other words, while the sniper who is less than 1,000 yards from his victim is not experiencing a condition of riskless warfare (since he can himself be killed by another enemy sniper, an artillery bombing, or by the naked soldier later the same day), the drone operator is completely safe. This debate first necessitates a discussion about the required criteria to determine when a soldier is no longer risking his life in a combat zone. Should it depend on the distance separating combatants and the reciprocal capacity to

---

[88] As Lionel McPherson rightfully wrote, 'unjust assaulters cannot fight back in the brute name of self-defence. The grounds for fighting matter' (2004, p. 491).

harm each other? In other words, when fighting enemy combatants armed only with AK47 and hand grenades, the principle of riskless warfare would apply to all soldiers who cannot be harmed with these weapons, which means operators of artillery pieces that can fire with accuracy up to dozens of kilometres away as well as aviators who cannot be shot down with these weapons. In return, this would mean that the members of the entity benefitting from this superior technology would need to forfeit using their weapons and use armaments similar to those of their enemy in order to have a legitimate right to kill. While we may think it is a matter of chivalry to only fire at technologically inferior enemy combatants when 'we see the whites in their eyes' to quote Israel Putnam (or some say William Prescott) in order to make it a 'fair fight', this thought is too ridiculous to entertain. Furthermore, in light of the nature of the terrorist threat, the drone operator remains a potential victim of the individual he is seeing on his screen. If he chooses not to launch his missile against him, that individual may directly or indirectly contribute to his death in the future. Indeed, this operator may be on board a commercial airliner that will be highjacked by the terrorist and crashed on a famous building; he may be in the vicinity of a WMD about to explode while taking a walk with his wife and children; or waiting for the next train in a metro station where that individual is planning to blow himself up or spread a nerve agent. Although this possibility is slim, it nonetheless exists and is not purely theoretical as it is precisely the *modus operandi* of terrorist organizations that no one ought to be safe from their harm.

Furthermore, the fact that they ought to be considered criminals and not soldiers implies that they can be targeted at any moment starting from when their threat to civilians meets the criterion of credibility. When this red line has been crossed, only their defection or permanent departure from the organization can allow them to regain their immunity against death[89]. By excluding from the understanding of pre-emption the notion of imminence, terrorists are not liable to being attacked only when they are bearing arms against their enemies but rather

---

[89] The latter point is problematic and may lead to what the Israeli Supreme Court has called the 'revolving door' problem as members of terrorist organizations may periodically leave and rejoin them as a way to regain their immunity. In order to prevent that loophole, the judges have determined that members of terrorist organizations are continuously liable to be attacked as they deemed that 'the rest[ing period they take] between hostilities is nothing other than preparation for the next hostility' (2006, sections 39–40). However, we cannot ignore the fact that individuals may be willing to leave these organizations on a permanent basis, although this option may end up being a death sentence as the organization may not tolerate this form of dissidence. This is why more subtle criteria ought to be considered in order to determine when individuals have really given up the terrorist path and have, as a consequence, regained their immunity against being killed.

as long as they are part of the organization posing a credible threat or as long as their organization continues to pose a credible threat. Indeed, limiting the notion of 'hostile acts' – which justifies defensive measures – to situations when terrorists are making use of their weapons or carrying them would simply bring us back to one of the options we ought to avoid, that is, states not having efficient tools to protect their civilians against terrorist attacks. From this perspective, hostility needs to be understood as a notion that covers not only the act itself but also its preparations from the earliest stages.

5.3 Who Can Be Targeted?

The last question that now needs to be answered is the identity of those who can legitimately be the targets of VATW. As said previously, members of terrorist organizations that pose a credible threat and who cannot be stopped with non-violent actions can rightfully be targeted through pre-emptive measures. In this regard, I have argued that drones can be a weapon of choice. But is it also justifiable to target individuals who are helping terrorists, such as members of the military or individuals involved in the production of WMD? In other words, were the killings of Iranian general Qasem Soleimani in January 2020 and of nearly half a dozen Iranian nuclear scientists since 2010 – with the most recent being the killing of Mohsen Fakhrizadeh in November 2020 – morally justifiable?

If the legitimacy of targeting with lethal means individuals who are the deciders, planners, and executioners of lethal threats against innocent people is not a matter to be questioned, the question of third parties who facilitate their murderous deeds deserves a more thorough analysis. What must first be discussed is their actual guilt and responsibility for the unlawful actions performed by others. I believe we can agree that they are not non-innocent: a notion that refers to individuals who either lack the prerequisites for moral accountability – such as children or individuals who are mentally impaired to a point where they are unable to distinguish between good and evil – or who are resorting to violence purely as a means of self-defence. Indeed, by providing or facilitating the acquisition of WMD by terrorists, by training them, and by refusing to cooperate with entities facing their threats, they become either accomplices or accessories to mass murder. There are, in this sense, no fundamental differences between these individuals and a man waiting outside the bank in his car while his friend robs it or someone hiding a wanted criminal in his apartment. This idea that individuals can be held accountable for their indirect participation in a crime applies not only to domestic laws but also to international law. For instance, this provision, which is part of Article 25(3)(d) of the Rome Statute, led to the conviction of Germain Katanga in 2004 for actions committed in the Democratic Republic of Congo[90]. However, in order to be considered an accomplice or an accessory to a crime, individuals' decisions to be indirectly involved in a criminal activity must result from a voluntary intent to do something unlawful. This is why individuals acting out of self-defence are considered innocent, as their intention to resort to lethal force is not the result of a voluntary decision to break the law.

Establishing the responsibility of individuals who are instrumental in the perpetration of a crime by a third party is only the first element of a more complex question. More precisely, if I am legitimately allowed to resort to deadly force against an armed bank robber who is posing a threat to my life, do I also have the right to resort to the same option against his accomplice waiting outside in the getaway car? Based on what has been said previously, unless that individual is also posing a deadly threat to me, it would not be necessary and proportional to harm him in a similar fashion as his comrade pointing a gun at me inside the bank. However, entities that have already been facing the terrorist threat for a long time have adopted a broader view of what constitutes 'direct participation' in a terror attack, such as the Israeli High Court that has concluded that this notion not only includes those who actively plan and perpetrate attacks but also those who collect intelligence, transport those who carry out the attacks and the weapons that will be used, as well as those who sell terrorists food or medicine, engage in general strategic analysis, supply general logistical support, or distribute propaganda (2006). Is this latter perspective justifiable?

I am afraid that I do not have a definite answer to that question. Obviously, embracing a very generous understanding of help and support is dangerous because it might end up resulting in depriving as many people as possible of their immunity against death. Since we can argue that food and water are as vital as weapons and ammunitions to terrorists, those bringing them these primary goods are as guilty as those providing them with WMD. The same could be said about a boat captain smuggling terrorists from the Middle East or North Africa to a location they are planning to attack in Europe. In order to prevent such a liberal interpretation, criteria need to be found by those defending the resort to

---

90 http://www.icc-cpi.int/en_menus/icc/press%20and%20media/press%20releases/Pages/pr986.aspx. With regard to Katanga's conviction for being an accessory to a war crime, the court found clear evidence that he had supplied guns to the militia that perpetrated the massacre for which he was accused, while also being fully aware that these weapons would be used against the civilian population and to commit other war crimes.

VATW against those collaborating with terrorists. For instance, similarly to what Michael N. Schmitt has argued, the 'criticality' of the help and support provided to those who are perpetrating direct harm ought to be a way to make that distinction (2004, p. 509). More precisely, this notion would imply distinguishing the types of actions that are essential to the perpetration of the crime from those that are not. But this idea probably comes with more questions than answers, with the most important being what exactly allows us to distinguish between critical and non-critical support. Schmitt only answers with an example that allows us to understand what he means without being precise about the required criterion that ought to be used[91]. Moreover, even if we were able to make that distinction, how would it be possible to determine with certainty that those who are deemed to have provided critical support acted in full knowledge that their help was critical to the terrorist attack? To use the previous example of the boat captain, how can we determine that he knew he was transporting terrorists to the location of their attack? How can we be sure that he did not act in order to prevent his wife and children from being killed by the terrorists? Acting under duress makes a big difference when it comes to determining the liability of individuals. The same is true for ignorance. If individuals who are in a situation of vincible ignorance can be blamed, it is not the same for those acting in a situation of invincible ignorance (for a thorough discussion of these notions, see Caron, 2019a). Making these assessments will invariably imply having a deep understanding of the individual's actions, which cannot be gained remotely. Thus, it is impossible in cases where gaining access to them is impossible or too dangerous; situations that, as it has been said, justify the resort to VATW. And this is precisely the problem, as the first criterion for justifying VATW would not be satisfied, since these people would be killed without us being able to determine their liability in the threat that is thought to be credible. For all these reasons, justifying VATW against the accomplices of terrorists or against those who are accessories to terrorist crimes seems to face insurmountable problems. Therefore, these violent measures short of war ought to be limited solely against those directly involved in the design, planning, and decision to strike against their targets.

---

91 He writes: 'For example, working in a munitions factory is distant from the direct application of force, whereas providing tactical intelligence is essential and immediate' (2004, p. 509).

Conclusion

Looking back at the last 20 years of the war on terror, it is difficult not to have a very critical assessment of its effectiveness and morality. With the withdrawal of American forces from Afghanistan, there are credible fears that the Taliban – who once harboured Al-Qaeda – may come back to power. The 2003 intervention in Iraq has, for its part, led to a destabilisation of the Middle East and to the birth of another terrorist organization – the Islamic State of Iraq and the Levant – that has happily followed the path of Al-Qaeda by picking up the torch of global terrorism. Twenty years after 9/11, it is, therefore, very difficult to pretend that the world is now safer. Furthermore, we cannot ignore that the unquestionable desire of Western nations to protect their citizens has been made possible at the expense of the lives of tens of thousands of innocent civilians living in the regions where our military actively fought these organizations. Paradoxically, the improper way in which we have fought terrorism tends to provide arguments to those supporting these groups, that we have ourselves acted in a terroristic manner. Thanks to this perspective, it is not hazardous to argue that resorting to war against terrorism was simply a largely inefficient and immoral strategy.

This is why this book has argued for the necessity to think of alternatives to war against these groups. The only problem is that the already accepted non-violent strategies associated with 'soft war' – such as arms and economic embargoes, diplomatic sanctions, naming and shaming, boycotts – are not necessarily efficient against terrorist organizations that are acting alone (they can also show their limits against states sponsoring and harbouring them). This is why there is a need to think of resorting to VATW in order for states targeted by the threat of terrorism to effectively protect their civilians in a way that will not be detrimental to the lives of innocent people who simply have the misfortune of living in territories where these criminal organizations are operating. These measures can be used as a means of reprisals or pre-emptively. If the first set of measures are already conventionally accepted by the international community, resorting to anticipatory measures is, for its part, more controversial. The main reason is that, because of the stateless nature of these groups, the criterion of imminence – which is a core element of the logic of the pre-emptive attack – simply does not apply to terrorism. Therefore, there is a need to find an alternative notion that will facilitate the justification of when it is legitimate to resort to force against these groups. Obviously, loosening up the rules surrounding this logic runs the risk of becoming a slippery slope that could ultimately erase the necessary distinction between pre-emption and prevention. This needs to be avoided and this is why I have suggested that the idea of terrorist groups posing what I

https://doi.org/10.1515/9783110729894-008

have defined as a 'credible threat' ought to be used. I leave to the reader the question of whether or not my argumentation was convincing. But the fact remains that the fight against the terroristic menace necessitates an in-between strategy that is neither entirely peaceful or non-violent nor entirely warlike. The international norm needs to evolve in light of this contemporary threat that was not a concern when the post-1945 Westphalian order was established, as it is now showing it limitations. The introduction of the R2P principle at the turn of the century as a response to the terrible massacres of innocent civilians in Rwanda, ex-Yugoslavia and elsewhere following the end of the Cold War has shown its capacity to face the new realities of our world. A similar path ought to be followed when it comes to the fight against global terrorism. Not doing so will leave the door open for states to employ disproportionate means of actions after they have suffered an attack or to the waging of preventive wars based on fallacious justifications.

Thus, there is no doubt in my mind that accepting VATW as legitimate solutions can provide states with additional means of action that can ultimately better allow the Just War Theory to deliver its promised goods: achieving just objectives by minimising as much as possible the effects of war and violence. After all, this theory, which I believe has been rightfully qualified by James Pattison as a pacifist approach,[92] makes it clear that war (understood in the way I have defined it in this book) ought to be a last resort option when all other solutions have failed or have proved their ineffectiveness. It is, however, problematic to define all forms of violence and usage of kinetic force as acts of war or akin to what Michael Gross and Tamar Meisels have called 'hard war'. As I have argued, violence is a complex notion that lies upon a spectrum of measures that cannot all be considered as acts of war, but rather as alternatives to war. In light of new threats and with a view to limiting the tragic consequences of war, it is time for the international community to acknowledge that it is at a turning point in its understanding of what is an acceptable use of force against terrorist groups. After all, and as Anne Frank once wrote, 'What is done cannot be undone, but at least one can keep it from happening again'. These words of wisdom should be at the heart of our assessment of how the Western world has dealt with terrorism since that fateful morning of 9/11.

---

92 As a contingent approach to peace and not as an absolute approach.

Bibliography

Almasy, Steve. 2019. "Dozens of people have been arrested over threats to commit mass attacks since the El Paso and Dayton shootings", CNN, August 22. https://edition.cnn.com/2019/08/21/us/mass-shooting-threats-tuesday/index.html

Aloyo, Eamon. 2015. "Just War Theory and the Last of Last Resort", *Ethics & International Affairs,* Vol. 29, No. 2, pp. 187 – 201.

Anscombe, G.E.M. 1981. *Ethics, Religion and Politics*, Vol. 3. Oxford: Basil Blackwell.

Aquinas, Thomas. *Summa Theologiae.*

Arafat, Yasser. 1974. "Speech at the United Nation General Assembly", November 13.

Arendt, Hannah. 1958. *The Origins of Totalitarianism.* Cleveland: World.

Augustine. *The City of God.*

Baldoli, Roberto. 2020. "Fighting Terrorism With Nonviolence: An Ideological Perspective", *Critical Studies on Terrorism,* Vol. 13, No. 3, pp. 464 – 484.

Bellamy, Alex. 2006. *Just Wars: From Cicero to Iraq.* Cambridge: Polity Press.

Beres, Louis René. 1991. "On Assassination Actual Armed Attack by Nonstate Actors", *The American Journal of International Law,* Vol. 106, pp. 770 – 777.

Bethke Elshtain, Jean. 2006. "Jean Bethke Elshtain Responds", *Dissent,* Summer, pp. 109 – 111.

Bethke Elshtain, Jean. 2013. "Prevention, preemption, and other conundrums". In Deen K. Chatterjee (ed.), *The Ethcis of Preventive War.* Cambridge: Cambridge University Press, pp. 15 – 26.

Bethlehem, Daniel. 2012. "Principles Relevant to the Scope of a State's Right of Self-Defense Against an Imminent or Actual Armed Attack by Nonstate Actors", *The American Journal of International Law,* Vol. 106, pp. 770 – 777.

Bin Laden, Osama. 1996. "Declaration of Jihad against Americans". https://www.911memorial.org/sites/default/files/inline-files/1996%20Osama%20bin%20Laden%27s%201996%20Fatwa%20against%20United%20States_0.pdf

Bin Laden, Osama. 2005. *Messages to the World: The Statements of Osama Bin Laden.* London: Verso.

Blake, Aaron. 2014. "Obama Says Islamic State 'Is Not Islamic'. Americans Disagree", *The Washington Post,* September 11.

Bloom, Mia M. 2004. "Palestinian Suicide Bombing: Public Support, Market Share, and Outbidding", *Political Science Quarterly,* Vol. 199, No. 1, pp. 61 – 88.

Boudreau, Donald G. 1993. "The Bombing of the Osirak Reactor", *International Journal on World Peace,* Vol. 10, No. 2, pp. 21 – 37.

Brennan, Maureen F. 1999. "Avoiding Anarchy: Bin Laden Terrorism, The U.S. Response, and the Role of Customary International Law", *Louisiana Law Review,* Vol. 59, No. 4, pp. 1195 – 1223.

Broad, William J., John Markoff and David E. Sanger. 2011. "Israeli Test on Worm Called Crucial in Iran Nuclear Delay", *New York Times,* January 15. https://www.nytimes.com/2011/01/16/world/middleeast/16stuxnet.html

Brown, Chris. 2013. "After 'Caroline': NSS 2002, Practical Judgment, and the Politics and Ethics of Preemption". In Deen K. Chatterjee (ed.), *The Ethics of Preventive War.* Cambridge: Cambridge University Press, pp. 27 – 45.

https://doi.org/10.1515/9783110729894-009

Brunstetter, Daniel and Megan Braun. 2013. "From *Jus ad Bellum* to *Jus ad Vim*: Recalibrating Our Understanding of the Moral Use of Force", *Ethics & International Affairs*, Vol. 27, No. 1, pp. 87–106.

Buchanan, Allen. 2013. "The Ethics of Revolution and Its Implications for the Ethics of Intervention", *Philosophy and Public Affairs*, Vol. 41, No. 4, pp. 292–323.

Buchanan, Allen and Robert Keohane. 2004. "The Preventive Use of Force: A Cosmopolitan Institutional Proposal", *Ethics & International Affairs*, Vol. 18, No. 1, pp. 1–22.

Bump, Philip. 2018. "15 years after the Iraq War began, the death toll is still murky", *The Washington Post*, March 20. https://www.washingtonpost.com/news/politics/wp/2018/03/20/15-years-after-it-began-the-death-toll-from-the-iraq-war-is-still-murky/

Bush, George W. 2002. U.S. National Security Strategy, *Prevent Our Enemies from Threatening Us, Our Allies, and Our Friends With Weapons of Mass Destruction*, West Point, New York, June 1.

Bush, George W. 2002. U.S. National Security Strategy, *Strengthen Alliances to Defeat Global Terrorism and Work to Prevent Attacks Against Us and Our Friends*, Washington, D.C., September 14.

Byman, Daniel L. 2009. "Do Targeted Killings Work?", The Brookings Institute, July. https://www.brookings.edu/opinions/do-targeted-killings-work-2/

Callimachi, Rukmini. 2015. "ISIS Enshrines a Theology of Rape", *New York Times*, August 13. https://www.nytimes.com/2015/08/14/world/middleeast/isis-enshrines-a-theology-of-rape.html

Caron, Jean-François. 2015. *La guerre juste: les enjeux éthiques de la guerre au 21ème siècle*. Québec: Les Presses de l'Université Laval.

Caron, Jean-François. 2018a. *A Theory of the Super Soldier: The Morality of Capacity-Increasing Technologies in the Military*. Manchester: Manchester University Press. *Théorie du super soldat: La moralité des technologies d'augmentation dans l'armée*. Québec: Les Presses de l'Université Laval, 2018.

Caron, Jean-François. 2018b. *Disobedience in the Military: Legal and Ethical Implications*. London: Palgrave MacMillan.

Caron, Jean-François. 2019a. *Contemporary Technologies and the Morality of Warfare: The War of the Machines*. London: Routledge.

Caron, Jean-François. 2019b. "Exploring the Extent of Ethical Disobedience Through the Lens of the Srebrenica and Rwanda Genocides: Can Soldiers Disobey Lawful Orders?", *Critical Military Studies*, Vol. 5, No. 1, 2019, pp. 1–20.

Caron, Jean-François. 2020. "Defining semi-autonomous, automated and autonomous weapon systems in order to understand their ethical challenges", *Digital War*, Vol. 1, No. 1–3, pp. 173–177.

Carter, David B. 2016. "Provocation and the Strategy of Terrorist and Guerrilla Attacks", *International Organizations*, Vol. 70, No. 1, pp. 133–173.

Cavallero, James, Stephan Sonnenberg and Sarah Knuckey. 2012. *Living Under Drones: Death, Injury and Trauma to Civilians From US Drone Practices in Pakistan*. New York: Stanford.

Chaliand, Gérard and Arnaud Blin. 2015. *Histoire du terrorisme: De l'Antiquité à Daesh*. Paris: Fayard.

Chamayou, Grégoire. 2015. *A Theory of the Drone*. New York: The New Press.

Chatterjee, Deen K. 2013. "Introduction". In Deen K. Chatterjee (ed.), *The Ethics of Preventive War*. Cambridge: Cambridge University Press, pp. 1–11.

Clough, David and Brian Stiltner. 2007. "On the Importance of a Drawn Sword: Christian Thinking about Preemptive War – and Its Modern Outworking", *Journal of the Society of Christian Ethics*, Vol. 27, No. 2, pp. 253–271.

Coady, C.A.J. 2008. *Morality and Political Violence*. Cambridge: Cambridge University Press.

Coady, C.A.J. 2013. "Preventive violence: war, terrorism, and humanitarian intervention". In Deen K. Chatterjee (ed.), *The Ethics of Preventive War*. Cambridge: Cambridge University Press, pp. 189–213.

Cole, Juan. 2015. "How 'Islamic' Is the Islamic State?", *The Nation*, February 24.

Connable, Ben, Jason H. Campbell and Dan Madden. 2016. *Stretching and Exploiting Thresholds for High-Order War*. Santa Monica: Rand Corporation.

Coolsaet, Rik. 2016. "Facing the Fourth Foreign Fighters Wave", *Egmont Paper* 81, March. https://www.jstor.org/stable/resrep06677.1?seq=1#metadata_info_tab_contents

Cronin, Audrey Kurth. 2002. "Behind the Curve: Globalization and International Terrorism", *International Security*, Vol. 27, No. 3, pp. 30–58.

Deakin, Stephen. 2014. "Naked Soldiers and the Principle of Discrimination", *Journal of Military Ethics*, Vol. 13, No. 4, pp. 320–330.

De Vattel, Emir. 2008. *The Law of Nations*. Indianapolis: Liberty Fund.

Dipert, Randall R. 2006. "Preventive War and the Epistemological Dimension of the Morality of War", *Journal of Military Ethics*, Vol. 5, No. 1, pp. 32–54.

Draper, Kai. 1998. "Self-Defence, Collective Obligation, and Noncombatant Liability", *Social Theory and Practice*, Vol. 24, No. 1, pp. 57–81.

Elagab, Omer Yousif. 1988. *The Legality of Non-Forcible Counter-Measures in International Law*. Oxford: Oxford University Press.

English, Richard. 2003. *Armed Struggle: The History of the IRA*. London: Palgrave MacMillan.

Ferwerda, Jeremy & Nicholas L. Miller. 2014. "Political Devolution and Resistant to Foreign Rule: A Natural Experiment", *American Political Science Review*, Vol. 108, No. 3, pp. 642–660.

Finlay, Christopher. 2015. *Terrorism and the Right to Resist: A Theory of Just Revolutionary War*. Cambridge: Cambridge University Press.

Fisk, Herstin and Jennifer M. Ramos. 2016. "Introduction". In Herstin Fisk and Jennifer M. Ramos (eds.), *Preventive Force: Drones, Targeted Killing, and the Transformation of Contemporary Warfare*. New York: New York University Press, pp. 1–29.

Fletcher, George. 1998. *Basic Concepts of Criminal Law*. Oxford: Oxford University Press.

Franck, Thomas M. 2002. Recourse to Force. State Action Against Threats and Armed Attacks. Cambridge: Cambridge University Press.

Franz, Nina. 2017. "Targeted Killing and Pattern-of-Life Analysis: Weaponized Media", *Media, Culture & Society*, Vol. 39, No. 1, pp. 111–121.

Freedman, Lawrence. 2003. *The Evolution of Nuclear Strategy*, 3rd edition. London: Palgrave MacMillan.

Freedman, Lawrence and Jeffrey Michaels. 2019. *The Evolution of Nuclear Strategy*. London: Palgrave MacMillan.

Friedrich, Carl and Zbigniew Brzezinski. 1965. *Totalitarian Dictatorship and Autocracy* Cambridge, MA: Harvard University Press.

Frowe, Helen. 2016. "On the Redundancy of Jus ad vim: A Response to Daniel Brunstetter and Megan Braun", *Ethics & International Affairs*, Vol. 30, No. 1, pp. 117–129.

Garren, David J. 2019. "Preventive Was: Shortcomings Classical and Contemporary", *Journal of Military Ethics*, Vol. 18, No. 3, 204–222.

Gartzke, Eric. 2013. "The Myth of Cyberwar: Bringing War in Cyberspace Back Down to Earth", *International Security*, Vol. 38, No. 2, pp. 41–73.

Gazette des Tribunaux. April 29, 1894, pp. 417–419.

George, Alexander L. 1991. *Forceful Persuasion: Coercive Diplomacy as an Alternative to War.* Washington, DC: United States Institute of Peace Press.

Glennon, Michael J. 2002. "Preempting Terorrism: The Case for Anticipatory Self-Defense", *The Weekly Standard*, 28 January, Vol. 7, No. 19.

Gollob, Sam and Michael E. O'Hanlon. 2020. *Iraq Index. Tracking Variables of Reconstruction and Security in Post-Saddam Hussein Iraq*, Foreign Policy at Brookings. https://www.brookings.edu/wp-content/uploads/2020/08/FP_20200825_iraq_index.pdf

Graham, David E., "Cyber Threats and the Law of War", *Journal of National Security Law & Policy*, Vol. 4, No. 1, 2010, pp. 87–102.

Gross, Emmanuel. 2006. *The Struggle of Democracy against Terrorism. Lessons from the United States, the United Kingdom, and Israel.* Charlottesville: University of Virginia Press.

Gross, Michael. 2010. *Moral Dilemmas of Modern War: torture, Assassination, and Blackmail in an Age of Asymmetric Conflict*, Cambridge: Cambridge University Press.

Gross, Michael. 2015. *The Ethics of Insurgency: A Critical Guide to Just Guerrilla Warfare.* Cambridge: Cambridge University Press.

Gross, Michael and Tamar Meisels (eds.). 2017. *Soft War: The Ethics of Unarmed Conflict.* Cambridge: Cambridge University Press.

Grotius, Hugo. 1913. *De Jure Belli ac Pacis.* Washington, D.C.: Carnegie Institute.

Harbom, Lotta and Peter Wallensteen. 2005. "Armed Conflicts and Its International Dimensions", *Journal of Peace Research*, Vol. 42, No. 5, pp. 623–635.

Heisbourg, François. 2001. *Hyperterrorisme: la nouvelle guerre.* Paris: Odile Jacob.

Henderson, J. Curtis. 1986. "Legality of Economic Sanctions Under International Law: The Case of Nicaragua", *Washington and Lee Law Review*, Vol. 43, No. 1, pp. 167–196.

High court of Israel. 2006. *Public Committee against Torture in Israel v. Government of Israel*, Case No. HCJ 769/02, December 13. http://elyon1.court.gov.il/files_eng/02/690/007/A34/02007690.a34.pdf

Hoffman, Bruce. 1993. "'Holy Terror': the Implications of Terrorism Motivated by a Religious Imperative", RAND Paper. https://www.rand.org/pubs/papers/P7834.html

Holmes, Robert L. 1989. *On War and Morality* Princeton, NJ: Princeton University Press.

Hopkinson, Michael. 2004. *The Irish War of Independence.* Dublin: Gill & MacMillan.

Hsu, Spencer S. 2015. "Judge orders Sudan, Iran to pay $75 million to family of USS Cole victim", *Washington Post,* March 31.

Huntington, Samuel. 1957. *The Soldier and the State: The Theory and Politics of Civil-Military Relations.* Cambridge, MA: Harvard University Press.

Irish Republican Army, General Headquarters. 1956. *Handbook for Volunteers of the Irish Republican Army. Notes on Guerrilla Warfare.* Boulder, CO: Panther Publications.

Iser, Matthias. 2017. "Beyond the Paradigm of Self-Defense? On Revolutionary Violence". In Samuel Rickless and Saba Bazargan (eds.), *The Ethics of War*, New York: Oxford University Press, pp. 207–226.

Josephus, Flavius. 1956. *Jewish War*, Vol. 1–2. London: William Heinemann Ltd & Cambridge, MA: Harvard University Press.

Juergensmeyer, Mark. 2001. *Terror in the Mind of God: The Global Rise of Religious Violence*. Berkeley: University of California Press.

Juvayni, Ata-Malek. 1958. *The History of the World-Conqueror*. Cambridge, MA: Harvard University Press.

Kahn, Paul W. 2002. "The Paradox of Riskless Warfare", *Faculty Scholarship Series*. Paper 326.

Kennan, George F. 1991. "Measures Short of War (Diplomatic)", in Giles D. Harlow and George C. Maerz (eds.), *Measures Short of War. The George F. Kennan Lectures at the National War College, 1946–47*. Washington, DC: National Defense University Press, pp. 3–20.

Keohane, Robert O. 2002. "The Globalization of Informal Violence, Theories of World Politics, and the 'Liberalism of Fear'", *Dialog-IO,* Spring, pp. 29–43.

Kilic v. Turkey. 2000. Council of Europe. European Court of Human Rights, March 28.

Laqueur, Walter. 1977. *Terrorism*. Boston: Little Brown.

Laqueur, Walter. 1987. *The Age of Terrorism*. Boston: Little Brown.

Laqueur, Walter. 1999. *The New Terrorism: Fanaticism and the Arms of Mass Destruction*. Oxford: Oxford University Press.

Laqueur, Walter. 2001. *A History of Terrorism*. New York: Little, Brown.

Laqueur, Walter. 2009. *Best of Times, worst of Times: Memoirs of a Political Education*. Waltham, MA: Brandeis University Press.

Luban, David. 2004. "Preventive War", *Philosophy and Public Affairs,* Vol. 32, No. 3, pp. 207–248.

Luban, David. 2007. "Preventive War and Human Rights". In Henry Shue and David Rodin (eds.), *Preemption: Military Action and Moral Justification*. Oxford: Oxford University Press, pp. 171–201.

Lubell, Noam. 2015. "The Problem of Imminence in an Uncertain World". In Marc Weller, Jake William Rylatt and Alexia Solomou (eds.), *The Oxford Handbook of the Use of Force in International Law*. Oxford Handbooks in Law. Oxford: Oxford University Press, pp. 697–719.

Lupton, Danielle and Valerie Morkevicius. 2019. "The Fog of War: Violence, Coercion and Jus ad Vim". In Jai Galliott (ed.), *Force Short of War in Modern Conflict: Jus ad Vim*. Edinburgh: Edinburg University Press, pp. 36–56.

Magenis, Sean D. 2002. "Natural Law as the Customary International Law of Self-Defense", *Boston University International Law Journal,* Vol. 20, no. 2, pp. 413–435.

Malanczuk, Peter. 1991. "The Kurdish Crisis and Allied Intervention in the Aftermath of the Second Gulf War", *European Journal of International Law,* Vol. 2, No. 2, pp. 114–132.

Marighella, Carlos. 1969. *Minimanual of the Urban Guerrilla*. https://www.marxists.org/archive/marighella-carlos/1969/06/minimanual-urban-guerrilla/

Martin, Brian. 2001. *Technology for Nonviolent Stuggle*. London: War Resisters' International, http://www.bmartin.cc/pubs/01tnvs/.

Martin, Brian. 2002. "Nonviolence Versus Terrorism", *Social Alternatives*, Vol. 21, No. 2, pp. 6–9.

McMahan, Jeff. 2004. "The Ethics of Killing in War", *Ethics*, Vol. 114, July, pp. 693–733.

McMahan, Jeff. 2006. "The Ethics of Killing in War", *Philosophia*, Vol. 34, pp. 23–41.

McPherson, Lionel K. 2004. "Innocence and Responsibility in War", *Canadian Journal of Philosophy*, Vol. 34, No. 4, pp. 485–506.

Metraux, Daniel A. 1995. "Religious Terrorism in Japan: The Fatal Appeal of Aum Shinrikyo", *Asian Survey*, Vol. 35, No. 12, pp. 1140–1154.

Miller, Seumas. 2009. *Terrorism and Counter-Terrorism: Ethics and Liberal Democracy.* Oxford: Blackwell.

Narveson, Jan. 1965. "Pacifism: A Philosophical Analysis", *Ethics*, Vol. 75, No. 4, pp. 259–271.

Nathanson, Stephen. 2010. *Terrorism and the Ethics of War.* Cambridge: Cambridge University Press.

Neff, Stephen C. 2005. *War and the Law of Nations. A General History.* Cambridge: Cambridge University Press.

Neumann, Peter R. 2009. *Old And New Terrorism.* Cambridge & Malden, MA: Polity Press.

Nicaragua case. 1986. Military and Paramilitary Activities in and Against Nicaragua (Nicaragua v. United State of America), ICJ Reports.

Nicholls, Steven. 2007. "Terrorism, Millenarianism, and death: A Study of Hezbollah and Aum Shinrikyo", Bachelor Thesis, Edith Cowan University. https://ro.ecu.edu.au/cgi/view content.cgi?article=2294&context=theses_hons

Nixon, Rob. 1992. "Apartheid on the Run: The South African Sport Boycott", *Transition*, Issue 58, pp. 68–88.

Office of the United Nations High Commissioner for Human Rights, Fact Sheet No. 32. https:// www.ohchr.org/documents/publications/factsheet32en.pdf

Orend, Brian. 2006. *The Morality of War.* Peterborough, Ontario: Broadview Press.

Orend, Brian. 2013. *The Morality of War*, 2nd edition. Toronto: Broadview Press.

Osman v. United Kingdom. 1998. Council of Europe. European Court of Human Rights, October 28.

Pasher, Yaron. 2015. *Holocaust Versus Wehrmacht: How Hitler's 'Final Solution' Undermined the German War Effort.* Lawrence, KS: University Press of Kansas.

Pattison, James. 2018. *The Alternatives to War: From Sanctions to Nonviolence.* Oxford: Oxford University Press.

Power, Samantha. 2002. *A Problem from Hell: America and the Age of Genocide.* New York: Basic Books.

Primoratz, Igor. 2013. *Terrorism: A Philosophical Investigation.* Cambridge, UK: Polity Press.

Proulx, Vincent-Joël. 2005. "Babysitting Terrorists: Should States be Strictly Liable for Failing to Prevent Transborder Attacks?", *Berkeley Journal of International Law*, Vol. 23, No. 3, pp. 615–668.

Raflik, Jenny. 2016. *Terrorisme et mondialisation: approches historiques.* Paris: Gallimard.

Rapoport, David C. 1984. "Fear and Trembling: Terrorism in Three Religious Traditions", *The American Political Science Review*, Vol. 78, No. 3, pp. 658–677.

Reike, Ruben. 2012. "Libya and the Responsibility to Protect: Lessons for the Prevention of Mass Atrocities", *St Anthony's International Review*, Vol. 8, No. 1, pp. 122–149.

Rockefeller, Mark L. 2004. "The Imminent Threat Requirement for the Use of Preemptive Military Force: Is It Time for a Non-Temporal Standard", *Denver Journal of International Law & Policy,* Vol. 33, No. 1, pp. 131–149.

Schachter, Oscar. 1984. "The Right of States to Use Armed Force", *Michigan Law Review,* Vol. 82, No. 5, pp. 1620–1646.

Schmid, Alex P. 1992. "The Response Problem as a Definition Problem", *Terrorism and Political Violence,* Vol. 4, No. 4, pp. 7–13.

Schmitt, Michael N. 2003. "The Sixteenth Waldemar A. Solf Lecture in International Law", *Military Law Review,* Vol. 176, pp. 364–421.

Schmitt, Michael N. 2004. "Direct Participation in Hostilities And 21st Century Armed Conflict". In Horst Fisher, Ulrike Froissart, Wolff Heinegg von Heintschel and Christian Rapp (eds.), *Crisis Management and Humanitarian Protection.* Berlin: BWV Berliner Wissenschafts-Verlag.

Schweller, Randall L. 1992. "Domestic Structure and Preventive War: Are Democracies More Pacific?", *World Politics,* Vol. 44, No. 2, pp. 235–269.

Sedgwick, Mark. 2004. "Al-Qaeda and the Nature of Religious Terrorism", *Terrorism and Political Violence,* Vol. 16, No. 4, pp. 795–814.

Sedgwick, Mark. 2012. "Jihadist Ideology, Western Counter-ideology, and the ABC Model", *Critical Studies on Terrorism,* Vol. 5, No. 3, pp. 359–372.

Shanahan, Timothy. 2009. *The Provisional Irish Republican Army and the Morality of Terrorism.* Edinburgh: Edinburgh University Press.

Shaw, Martin. 2005. *The New Western Way of War.* Cambridge: Polity Press.

Shelah Shehadeh Special Investigatory Commission. 2011. https://mfa.gov.il/MFA/Abou tIsrael/State/Law/Pages/Salah_Shehadeh-Special_Investigatory_Commission_27-Feb-2011.aspx

Shue, Henry. 1996. *Basic Rights: Subsistence, Affluence, and US Foreign Policy,* 2nd edition. Princeton: Princeton University Press.

Singer, David J. and Mel Small. 1972. *The Wages of War, 1816–1965: A Statistical Handbook.* New York: Wiley.

Smith, J. Warren. 2007. "Augustine and the Limits of Preemptive and Preventive War", *Journal of Religious Ethics,* Vol. 35, No. 1, pp. 141–162.

Smyth, H.D. 1945. *A General Account of the Development of Methods of Using Atomic Energy for Military Purposes under the Auspices of the United States Government 1940–1945.* Washington, D.C.: USGPO, August.

Stiennon, Richard. 2015. "A Short History of Cyber Warfare". In James A. Green (ed.), *Cyber Warfare: A Multidisciplinary Analysis.* Abingdon, Oxfordshire: Routledge, pp. 7–32.

Strachan, Hew. 2007. "Preemption and Prevention in Historical Perspective". In Henry Shue and David Rodin (eds.), *Preemption: Military Action and Moral Justification.* Oxford: Oxford University Press, pp. 23–39.

Strawser, Bradley Jay. 2014. *Killing bin Laden: A Moral Analysis.* New York: Palgrave MacMillan.

Strehle, Stephen. 2004. "Saddam Hussein, Islam, and Just War Theory: The Case for a Preemptive Strike", *Political Theology,* Vol. 5, No. 1, pp. 76–101.

Svarc, Dominika. 2006. "Redefining Imminence: The Use of Force Against Threats and Armed Attacks in the Twenty-First Century", *Journal of International & Comparative Law,* Vol. 13, No. 1, pp. 171–191.

Taylor Wilkins, Burleigh. 1992. *Terrorism and Collective Responsibility.* London: Routledge.

The 9/11 Commission Report. https://avalon.law.yale.edu/sept11/911Report.pdf

Thiessen, Marc A. 2011. "Iran Responsible for 1998 U.S. Embassy Bombings", *Washington Post,* December 8.

Thompson, Allan (ed.). 2007. *The Medias and the Rwanda Genocide.* Ann Arbor: Pluto Press.

Trachtenberg, Marc. 2007. "Preventive War and US Foreign Policy". In Henry Shue and David Rodin (eds.), *Preemption: Military Action and Moral Justification.* Oxford: Oxford University Press, pp. 40–68.

Uniacke, Susan. 2007. "On Getting One's retaliation in First". In Henry Shue and David Rodin (eds.), *Preemption: Military Action and Moral Justification.* Oxford: Oxford University Press, pp. 69–88.

US Department of State. Glossary. https://2001-2009.state.gov/s/ct/info/c16718.htm

Velásquez Rodríguez Case, Inter-Am.Ct.H.R. (Ser. C) No. 4 (1988), Inter-American Court of Human Rights (IACrtHR), 29 July 1988. https://www.refworld.org/cases,IACRTHR,40279a9e4.html

Voltaire. 1764. "War". In *The Works of Voltaire, Vol. VII (Philosophical Dictionary Part 5).* https://oll.libertyfund.org/titles/voltaire-the-works-of-voltaire-vol-vii-philosophical-dictionary-part-5

Von Clausewitz, Carl. 1976. *On War.* Translated by Michael Howard and Peter Paret. Princeton, NJ: Princeton University Press.

Walzer, Michael. 2002. "No Strikes", *The New Republic.* 30 September. https://newrepublic.com/article/61959/no-strikes

Walzer, Michael. 2004. *Arguing About War.* New Haven & London: Yale University Press.

Walzer, Michael. 2006. *Just and Unjust Wars: A Moral Argument with Historical Illustrations,* 4th edition. New York: Basic Books.

Watson Institute. 2020. https://watson.brown.edu/costsofwar/costs/human/civilians/afghan

Weinberg, Leonard, Ami Pedahzur and Sivan Hirsch-Hoefler. 2004. "The Challenges of Conceptualizing Terrorism", *Terrorism and Political Violence,* Vol. 16, No. 4, pp. 777–794.

William Eduardo Delgado Páez v. Colombia, Communication No. 195/1985, U. N. Doc. CCPR/C/39/D/195/1985 (1990).

Wolfendale, Jessica. 2017. "Defining War". In Michael L. Gross and Tamar Meisels (eds.), *Soft War: The Ethics of Unarmed Conflict.* Cambridge: Cambridge University Press, pp. 16–32.

Yoo, John. 2003. "International Law and the War in Iraq", *The American Journal of International Law,* Vol. 97, No. 3, pp. 563–576.

Zenko, Micah. 2017. "Obama's Final Drone Strike Data", Center for Preventive Action. January 20. https://www.cfr.org/blog/obamas-final-drone-strike-data

Index

https://doi.org/10.1515/9783110729894-010